Integrated Watershed Management

Connecting People to Their Land and Water

Integrated Watershed Management

Connecting People to Their Land and Water

Hans M. Gregersen
Peter F. Ffolliott
Kenneth N. Brooks

www.cabi.org

CABI is a trading name of CAB International

CABI Head Office
Nosworthy Way
Wallingford
Oxfordshire OX10 8DE
UK

Tel: +44 (0)1491 832111
Fax: +44 (0)1491 833508
E-mail: cabi@cabi.org
Web site: www.cabi.org

CABI North American Office
875 Massachusetts Avenue
7th Floor
Cambridge, MA 02139
USA

Tel: +1 617 395 4056
Fax: +1 617 354 6875
E-mail: cabi-nao@cabi.org

A catalogue record for this book is available from the British Library, London, UK.

A catalogue record for this book is available from the Library of Congress, Washington, DC.

Library of Congress Cataloging-in-Publication Data
Gregersen, H.M.
   Integrated watershed management : connecting people to their land and water / Hans M. Gregersen, Peter F. Ffolliott, Kenneth N. Brooks.
       p.   cm.
   Includes bibliographical references and index.
   ISBN 978-1-84593-281-7 (alk. paper) -- ISBN 978-1-84593-282-4 (ebook)
1. Watershed management. 2. Water quality management. 3. Land use.
4. Conservation of natural resources. I. Ffolliott, Peter F. II. Brooks, Kenneth N.
III. Title.

TC409.G684 2007
333.7--dc22

2007007403

ISBN-13: 978 1 84593 281 7

Typeset by SPi, Pondicherry, India.
Printed and bound in the UK by Cambridge University Press, Cambridge.

Contents

Preface and Overview

The challenge of managing the environments in which people live becomes more complex as human populations expand, demands on the natural resources increase and new technologies are developed that let people – knowingly or unknowingly – destroy their environment with less effort than in the past. People have learned through experience that what we do to upstream watersheds can drastically affect large numbers of people and their lands and water downstream. We have also learned that what appear to be isolated human interventions interact with each other – and sometimes the effects amplify each other and can in the aggregate affect humankind across vast landscapes, river basins or even at the global scale. The World Commission on Sustainable Development's dictum – 'think globally, and act locally' – thus has become increasingly clear, relevant and urgent.

At the same time that the challenges for land and water management increase, there are opportunities to halt the degradation and destruction and, in doing so, create more sustainable environments in which people can live and satisfy their basic needs. However, while many of these opportunities might exist on paper, the reality is that local political units – different states, provinces, towns, cities, villages and communities – determine what happens within their jurisdictions. Almost invariably these jurisdictions are situated within, but are not coincident with, the boundaries of a watershed or a river basin. Nevertheless, while water flows downstream, many of the natural processes associated with the flowing water occur without regard for political boundaries. This means that for the opportunities for sustainable land and water husbandry to be realized, political units and civil society must come together to cooperate and think broadly and collectively. Institutional mechanisms need to be developed to create incentives for action across political boundaries focused on the common good within a river basin or watershed. But the institutional mechanisms can only be effectively used if they are based in the technical realities of what is going on with the soil, water and other natural resources and the interactions between these resources.

Many organizations and formal and informal groups guide and affect the use and management of land and water resources on most watersheds. *This book is intended for people involved in these entities*, including government agencies, quasi-governmental groups such as river basin commissions, civil society-dominated water user groups, watershed associations, councils, partnerships, non-governmental organizations (NGOs) and other private groups that might have some local government input as one of many partners. In the USA alone, there are thousands of such local and regional groups. The directory of the River Network, a unique national organization dedicated to supporting grass-roots groups working for watershed protection, lists more than 4000 groups (available at: http://www.rivernetwork.org). As discussed in Chapter 3, there is a rapid growth in public involvement in watershed management in many countries across the globe. There is global recognition of the issues and the opportunities for grass-roots action.

This book provides background information and presents factors to be considered and procedures that facilitate organizing and guiding land and water use in concert with one another; it further serves as a reference for planning, monitoring and implementing development efforts and natural resource management through the lens of integrated watershed management (IWM). A premise of this book is that an effective interaction or combination of institutional mechanisms and technical information is essential for successful IWM that results in lasting benefits to the people (stakeholders) living on the watershed or the river basin.

Effective approaches have been developed to ensure that people in a watershed have a common and positive overall goal and plan for the use of land, water and other natural resources and that their actions do not adversely affect future generations. While these approaches might differ in detail, they all have some of the common elements that comprise IWM – the theme of this book.

The many *challenges and opportunities* for improving people's use and management of their land and water resources are presented in Chapter 1. The loss of protective forest cover in upland watersheds, urban encroachment on the landscape, loss of wetlands, pollution of water supplies and the accelerated loss of soil resources represent growing challenges. The opportunities to cope with these issues and to develop sustainable solutions to problems and issues of land and natural resource scarcity, water scarcity and water quality, the paramount issues faced in the 21st century, are introduced in this chapter.

Land use, watershed management and cumulative effects are the subjects of Chapter 2. People want many goods and services from the watersheds on which they live, including water for personal consumption, irrigation and hydropower; raw materials for wood products; agricultural crops, meat and dairy products; recreation and tourism; and maintenance of biodiversity. A fundamental objective of IWM is developing, managing and sustaining production systems that are suited to the existing environment and natural resources base and that can be sustained for future generations. Likewise, an objective of IWM is sustaining, and in some instances increasing, supplies of high-quality water and improving the quality of water in streams, lakes and other water bodies. Preventing excessive soil erosion is fundamental to IWM.

What distinguishes the IWM approach from other land management practices is its holistic consideration of the linkages among all such objectives and the activities within a watershed aimed at meeting them. Accepting the reality of the fact that the watershed is made up of a collection of land-use units of varying physical sizes, each of which might be defined by different ownership or control, we consider the interactions among such units and the activities undertaken on them including agricultural cropping, livestock production, forestry and agroforestry activities, infrastructure development and urbanization.

We make the point that all the users of land, water and natural resources should recognize that they need to play the role of *watershed manager*. Although there are few with the formal title of watershed manager, people need to understand the interactions of their use of land and water within the context of watershed boundaries. This requires that we become aware of upstream–downstream linkages and what we call *cumulative effects*. Multiple land-use activities across a watershed can bring about a variety of changes that might not seem important when viewed in isolation. However, in the aggregate, they can affect the quantity and quality of water flowing from upland watersheds and impact on people living downstream.

The *institutional context* of IWM is the subject of Chapter 3. Many factors are converging to cause natural resource managers, researchers, decision makers and society as a whole to look increasingly to IWM as a participatory and practical approach for addressing land, water and other natural resources problems. Avoidance of many potentially serious problems depends largely on moving ahead with programmes of IWM that prevent the degradation of natural resources and build up the capacity to use resources wisely. The institutional context for such programmes is critical. The more society and governments accept and promote IWM, the more likely that there will be changes in the relationships between the public and private institutions that determine the governance of land, water and other natural resources.

Innovative institutional mechanisms exist to guide activities on a watershed. A complex set of intertwined laws and customary rights affect a watershed and the water that flows through and from it in most cases. Laws, regulations and policies exist in some instances to help determine the effectiveness with which the responses to increasing demands and needs for water and other natural resources are implemented. They are implemented through the promotion of local commitment and participation; legal and regulatory mechanisms; fiscal and financial mechanisms creating incentives that influence private behaviour; and public investment and improved management of resources. A combination of these mechanisms often is the most effective approach to policy implementation, including customary rights, laws creating statutory rights, treaties and incentives and mechanisms that create effective markets for water and other natural resources.

The use of economic instruments in dealing with institutional issues relating to IWM is increasing but has far from reached its full potential. Economic tools can offer several advantages such as providing incentives to change behaviour; raising revenue to help finance necessary investments; and/or establishing user priorities to attain management objectives at the least possible overall cost to society. Designing appropriate economic instruments requires considerations of efficiency, environmental sustainability, equity and other social concerns and the complementary institutional and regulatory framework. These economic instruments include water prices, tariffs and subsidies, incentives, fees and fee structures, water markets, privatization and taxes.

Planning and policies for IWM are discussed in Chapter 4. Planning is an orderly process that people use to understand and deal with future trade-offs, uncertainties and complexities that hinder clear and non-controversial decisions and actions. Planning is necessary because there are likely to be trade-offs between different interests and priorities among stakeholder groups in a watershed. Furthermore, there are less easily defined intergenerational trade-offs – what we do today in a watershed or a river basin affects the options available to future generations. As indicated in this chapter, planning is essential for achieving the most effective, efficient and equitable results possible. Because the planning process involves working with trade-offs in most IWM processes, the key here is to plan to achieve an equitable and acceptable balance between the interests of upstream and downstream stakeholders and between the different political units within the watershed or river basin. While most often planning is political in nature, it should involve technical input related to knowledge of hydrology, land management, planning methods, economics and the means to use this input to reduce risk to acceptable levels.

The general elements in the planning process for IWM are many. However, the two main inputs to this process are technical inputs on the one hand and political negotiation on the other. If the two sets of inputs interact effectively in the planning process, the result hopefully is an informed and fair set of planning guidelines on the actions needed by all key stakeholders to move ahead towards a common set of goals. Within this planning context, Chapter 4 explores the lessons that have been learned from past planning exercises around the world. Deriving from past experience and the lessons discussed above, we identify a set of practical and workable steps in IWM planning efforts.

Chapter 5, *hydrological processes and technical aspects*, provides the background that helps people understand the hydrologic response of watersheds to land use and management. The hydrological effects of land use and watershed management activities are manifested by changing water flow regimes, altered erosion–sedimentation relationships, changes in productivity of the land and changes in water quality and aquatic ecosystems. An understanding of these effects requires knowledge of hydrology within an interdisciplinary perspective. The climatic regime and the vegetative, geologic, topographic and soil characteristics of a watershed determine how the hydrologic processes of the watershed respond to precipitation inputs. Hydrologic processes on watersheds that undergo land-use activities are influenced principally through changes in the character of the vegetative cover, the physical and biological characteristics of soils and changes in watershed features such as stream channels, riparian corridors, lakes, ponds and reservoirs. The extent and persistence of anthropogenic changes such as urbanization influence the magnitude and persistence of the hydrologic response. Many of the environmental issues of concern in the 21st century encompass questions of various scales that will only be answered by improving the status of our knowledge of hydrologic processes.

Monitoring and evaluation to improve understanding and performance are dealt with in Chapter 6. The linkages among watershed management, changes in environmental systems and the welfare of people are not always well understood. Although we plan for particular welfare effects when an IWM activity is undertaken, we cannot always be sure that the activity will be implemented as planned or that the activity will have the effects on natural resources, the environment and welfare of people as anticipated. In responding to this problem, monitoring and evaluation (M&E) systems become key elements in the planning and implementation of activities within the IWM framework. What is unique about M&E in this context is that we consider both the on-site consequences of practices on the land and the downstream off-site effects (externalities) due to the practice, project or programme. Without effective M&E, people generally have only a little basis to determine the effects of land- and water-use activities on a natural resource base, the environment and their welfare. People also need M&E to facilitate changes in policies, practices and other activities that are required to meet their desired goals and provide the feedback that can help improve performance of future watershed management practices.

While the people responsible for M&E cannot insure that the information they produce will be used in future planning, managing and setting of policies, they can target their efforts to meet effectively and efficiently the information needs of the users. Failure to involve the users in the design of M&E efforts; failure to obtain the needed information and incorrect or incomplete data analysis; failure to provide the information in a form that can be readily used; and failure to provide the information when and where it is needed will almost certainly ensure that the resulting information will not be used. The best ways to insure that this does not happen are discussed in this chapter.

Chapter 7 discusses *research lessons learned and knowledge transfer*. Watershed managers recognize the purpose of research in helping to provide the information needed for the formulation of better IWM practices. The basic purpose is meeting the longer-term and broader information needs of people involved with IWM – the watershed managers. We use the term watershed managers to include those with formal titles and also all the other land and water users who think of themselves in terms of a holistic management context. Only by understanding their respective roles can managers and researchers work together in satisfying people's expectations for their land, water and other natural resources.

If IWM processes are going to be successful, people need to be participatory and to develop coordination and cooperation among all the stakeholders in a watershed or a river basin. This means that all stakeholders need a minimum level of knowledge of objectives, goals and means, and specific information related to their specific roles. Research findings and management experience provide the information and lessons needed; training and education provide the means for effectively transferring that knowledge where it needs to go and when it is needed. As this chapter points out, there are many training and knowledge transfer options available to the planners and managers of an IWM process and programme, and it is therefore important to pick the appropriate training tool to meet the goals and objectives involved.

Chapter 8 summarizes the role of IWM in dealing with the issues of today and presents ideas on how we can go about using the IWM approach to cope with the uncertainties of tomorrow. Adaptive management offers an approach for planners, and watershed managers to deal with cumulative effects across watershed landscapes that are subjected to changing demographics and changes in climates. It also provides a means to take maximum advantage of the lessons we learn from past mistakes and successes and to introduce such lessons into current and future practice.

Annexes provide some of the detailed technical underpinnings that we feel will be helpful in better understanding the multidisciplinary content in this book and in accessing more in-depth understanding of the various topics covered. We fully realize that in an overview guide such as the present one it is neither possible nor desirable to cover all the details for each element in the overall framework.

1 Challenges and Opportunities

The challenge of managing the environments in which people live becomes more complex and difficult as human populations increase, demands on the natural resource base increase and new technologies are developed that let people, willingly or unknowingly, destroy their environment at a more rapid rate and with less effort than in the past. We have a greater capacity to manipulate land and water for our benefit, but we also have a greater capacity to damage or destroy our environment. The expanding pollution of the world's water and air resources, the large-scale deforestation taking place, the destruction of fisheries, the drawdown of water tables and the depletion of key natural resources on which people depend for their livelihood are ample evidence.

At the same time that these challenges increase, the opportunities to halt the degradation and destruction and create more sustainable and dynamic environments in which people can live and satisfy their needs for natural resources, space and enjoyment of nature also increase. New knowledge is being created daily from research efforts and management experiences on how the environment works, how natural resources interact and how people obtain what they need in more environment-friendly ways. New information and communication technologies (ICTs) also are being developed that permit people to place the new knowledge into practice more rapidly and more broadly. New understanding of why and how people destroy their environment is being translated into plans and actions to prevent or reduce such destruction. More effective regulatory and incentive mechanisms are being developed, tested and put to use. The successes and failures of land and water management and use are shared more easily and rapidly among nations through the use of the new ICTs such as the Internet and its many technology-transfer applications (see Chapter 7). Achieving more sustainable means of increasing food security and reducing poverty are fundamental issues related directly to reducing environmental degradation in much of the developing regions. People in the wealthier, more highly developed countries are faced with difficult choices of sustaining or increasing food, water and other resources while proactively protecting ecosystems, their diversity and productivity. To do so requires that we manage land and water in concert with one another.

There is also a rapidly increasing recognition that the land and water management activities within watershed landscapes can produce a number of national, regional and sometimes global public goods, and that there is reason for the people of the world to be interested in paying local watershed management stakeholders for such environmental services. The application of payments for environmental services is growing. For example, the Clean Development Mechanism (CDM), under the Kyoto Protocol on climate change, includes land and water management and hydropower projects that generate carbon credits within the guidelines of the Kyoto

Protocol of the United Nations Framework Convention on Climate Change. One specific example is the 155MW La Higuera hydropower project in Chile. It is expected to lead to an estimated 470,000 t of CO_2 emission reductions and carbon credits per year, reductions for which others are willing to pay (EcoSecurities, 2006). In the case of these projects, watershed management activities needed to sustain and stabilize the flows of water that feed the hydropower projects can be funded through the payments by outsiders for carbon credits. Similar ideas have been put forth to finance large-scale wetland restoration on the basis that wetlands naturally remove nutrients in the Mississippi River Basin, thereby reducing nutrient loading and hypoxia in the Gulf of Mexico. Donald Hey (Wetlands Initiative, 2003) has developed a strategy in which landowners restore and manage wetlands to harvest nutrients from water, selling credits of the removed nutrients to wastewater treatment plants and others who are required to discharge cleaner water by law.

Increasingly with the above kinds of developments and broadening interactions in mind, the logic of The World Commission on Sustainable Development statement to think globally and act locally has become clear. We have learned that what appear to be isolated human interventions in nature interact with each other and, sometimes, the effects amplify each other and can in the aggregate affect humankind at the global scale. The innocent-looking pollutants we release into the air come back to haunt us by, for example, causing global change in weather patterns and increased respiratory problems. What we do to the land upstream in a watershed or river basin can drastically affect large numbers of people and their land and water downstream in a river basin or watershed.

The distinction made between a river basin and a watershed differs depending on one's viewpoint. Here, we refer to a *river basin* as a large unit of land that drains into an ocean. The term *watershed* is used to refer to smaller units that contain all lands and waterways that drain to a given common point. A river basin can, therefore, contain many watersheds within its boundaries.

We need to think broadly about problems and their interrelated causes and interactions to take advantage of opportunities for positive interactions and to control the negative ones. In some cases, this means thinking nationally, regionally or globally, while in other cases it means thinking at the river basin or watershed level. In all cases, it means acting locally as well as regionally, and globally in some cases. It turns out that most global environment-related issues can only be solved, ultimately, by dealing on the ground at local levels that are defined by existing physical boundaries and controlled by institutions. A national government can make a decision on a solution to problems such as soil degradation, water quality improvement or deforestation, but as we know from many examples in the real world, the solutions only work if people act locally.

The challenge then is to move from broad thinking and decisions to effective local actions. It is no different at either the watershed or river basin level, because of the biophysical and socio-economic interactions and interrelationships within the land units. We must therefore plan on a broad basis that includes the whole river basin and the many interactions that take place within its boundaries. These interactions result because of the two common physical threads for all watersheds:

- Water flows downstream ignoring all political boundaries en route.
- Most of the things that people do to their land and water upstream affect the water quantity, timing of flow and quality downstream and, as a consequence, downstream land productivity in its various forms.

While these interactions can be recognized and taken into account for the entire river basin at the planning level, the reality of the world is that different and independent local political units – states, provinces, towns, cities, villages and communities – are almost invariably situated within the boundaries of a river basin or cut across the boundaries of the watersheds contained in the river basin. Furthermore, each of these political units acts on its own to serve its self-interest in the absence of incentives or laws to create coordinating mechanisms. These interactions give rise to a fundamental challenge: while we can think broadly, how can we translate the conclusions and solutions derived at that broad, often conceptual, level into effective, complementary local actions? Being able to meet this challenge depends largely on the institutional mechanisms that can be developed to create incentives for action focused on the common good. However, and importantly, the institutional mechanisms can only be effective if they are grounded in the technical realities of what is going on with the soil, water and biophysical resources they are focused on and the interactions between these resources. Thus, a fundamental premise of this book is that an effective interaction or combination of institutional and technical information is required for successful watershed management that results in lasting benefits to the stakeholders living in the watershed or river basin.

Fundamental Questions Addressed

With the above in mind, the two fundamental questions addressed in this book are:

1. What mechanisms exist to ensure that people in a watershed have a common and positive overall goal of land and water use and that their actions do not adversely affect land and water resources for future generations?
2. How do we enable stakeholders to act in a cooperative and coordinated fashion to achieve these goals as best possible?

These questions can only be answered if we have good knowledge of the biophysical realities of the watershed and the biophysical interactions that take place within it in response to natural forces and the actions of humans. We also need an understanding of the institutional dynamics of the communities that exist within its boundaries, the interactions that exist between them at present and the motivations of the various stakeholders and the incentives that would influence them to change their motivations and actions that adversely affect the watersheds within which they live and work.

Over the years, effective approaches have been developed to generate the needed information to answer these questions and implement the actions implied by the answers. While the approaches might differ in detail, they all have some of the common elements which make up the main theme of this book. Some people call the chosen approach integrated watershed management (IWM), integrated catchment management or integrated river basin management. Others refer to it as integrated natural resources management or integrated water resource management when the focus is more directly on the water resources involved. The approach in the European Union is called the *ecosystem approach in water management* or more simply *ecosystem management* (ECE, 2004), which is described as

> the idea that water resources should not be managed in isolation from other ecosystem components, such as land, air, living resources and humans present in the watershed. The watershed is thus considered as an entire ecosystem. The protection,

sustainable use and restoration of its components are essential for the sustainability of water resources management.

We use the term IWM in this book, recognizing that all approaches and terms embrace the same ideas when it comes to fundamental principles. The overall watershed management relationships considered in this book are illustrated in Fig. 1.1. We consider some of the challenges faced and opportunities for overcoming these challenges in the following pages.

Challenges Faced

Sustaining or improving people's livelihood and their general well-being is the main purpose of the IWM approach to land stewardship everywhere and is the main challenge faced by those working with watershed management. Increasing human populations and the need of alleviating or preventing water scarcity will remain one of the, if not the paramount, natural resource–related issue faced in the 21st century. Continued loss of protective vegetative covers on upland watersheds and (in turn) the accelerated loss of soil resources represents a continuing challenge, especially soil losses and nutrient depletions in the humid tropical regions of the world. But there are also many other challenges, as detailed in the following pages with examples.

The challenges of managing land, water and people within the IWM framework are created or affected by the social and economic dynamics that exist and affect the environment of the watershed. Overcoming the challenges is becoming an increasingly difficult task to achieve with the world's increasing human populations that put added pressures on the land and water resources. Worldwide, the ratio of land to population is dwindling, which means increased pressures on the land. The amount of cultivated agricultural land is less than 0.5 ha for each person in many countries. Food production has been increasing in many countries by cultivating additional lands or extending livestock-grazing areas. Now, these options are rapidly disappearing. More than 6 billion human beings presently inhabit the earth. Each day, 200,000 more individuals are added to our planet's human population. As pressures on the world's natural resources mount, not only because of the growth of human populations but also because of economic growth, larger per capita demands are placed on natural resources. Concurrently, the need increases for even more intensive and careful stewardship of the natural resource base on which all humankind depends for survival.

The challenge in meeting the needs for natural resources by increasing human populations is illustrated in the mountains of Nepal. The increasing human population of Nepal and its neighbouring countries in this densely populated part of the world are continually moving farther into the mountains and higher up the slopes to seek a means of bettering their livelihood. Even with the aid of terracing and other erosion-prevention measures, these slopes are too steep and the soils too thin to sustain intensive agricultural cultivation. Demands of the increasing human populations result in the cultivation of less suitable soils and steeper lands, causing a reduction of productivity per hectare of land. One hectare of cultivated land must support ten or more people in many parts of Nepal. As much as 40% of what was once productive agriculture in the eastern hills has been abandoned by farmers and allowed to revert to mostly scattered shrubs because it is no longer fertile enough to support sustainable agriculture. These lands are also the sites and sources of excessive soil erosion, accelerated

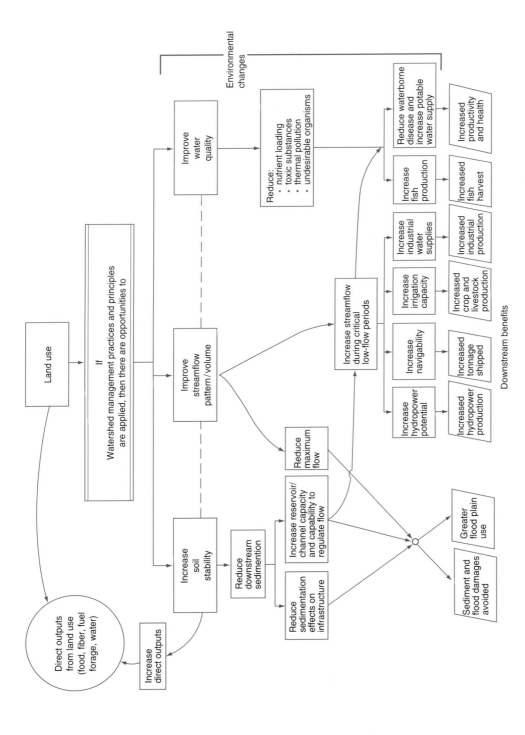

Fig. 1.1. Relationships between physical effects, environmental changes and downstream benefits from watershed management practices as compared to the 'without' practices condition. (From Brooks *et al.*, 1990, modified from Gregersen *et al.*, 1987.)

gully erosion and massive landslides. However, agricultural cultivation is only partially responsible for the rapid deterioration of the watershed landscapes. Status of the remaining forests in Nepal continues to be in jeopardy because of overgrazing and harvesting fodder for the increasing numbers of livestock. Forest and rangeland fires also add to the problem. The problems of Nepal are repeated in many countries.

Water scarcity

Water scarcity continues to be a global issue that grows more serious with the ever-expanding human populations. Often thought to be a problem only in arid countries, the dire implications of water scarcity have far-reaching consequences to people in almost every region of the world. (*Water scarcity* is a relative term, since more water can often be obtained but at a high cost such as desalinization of seawater, a cost that many cannot afford.) Problems of water scarcity are caused by a variety of reasons including the growing demands for water in the face of limited supplies, inefficient use of existing water supplies, ineffective and sometimes inequitable distribution of water to users and pollution problems that make available water supplies unusable. High population growth rates continue to increase the urgency of dealing with these causes of water scarcity. Solutions are largely impeded by the many conflicting interests in water and ineffective efforts to attain solutions to water scarcity. The poor are often the people hurt the most by water scarcity.

Increasing people's interest in the conservation and protection of limited water supplies is a key to solving the problems of scarce water supplies. In terms of IWM, relationships between management practices on upstream watersheds and the resulting effects of these management practices on the quantity, quality and delivery of water to downstream users are often a part of this solution. As mentioned above, the flow of water follows watershed boundaries, not political boundaries. Therefore, what is done on upland watersheds of one country, state, community or landowner can significantly affect other countries, states, communities or landowners occupying a downstream location (Brooks *et al.*, 1994). This has been the case within large river basins such as the Mekong River Basin, the Amazon River Basin, the Congo and Nile river basins and the Mississippi River Basin. Recognizing the off-site effects of watershed management practices in the efficient transport of high-quality water supplies in the amounts needed to people in downstream situations is therefore necessary for developers and managers of water and land resources.

Loss of protective vegetative cover and accelerated loss of soil resources

Another challenge to the management of the interactions between people and their land and water resources is overcoming the continuing loss of the protective vegetative cover critical to the prevention of accelerated loss of soil resources on fragile watershed lands. This protective vegetative cover is the principal means of controlling excessive overland water flows and soil erosion. One impact of the loss of soils has been increased flooding of valleys and shifting of streambeds causing more water and silt to invade prime agricultural land, irrigation and hydropower structures, human settlements and communication systems. During the dry periods of the year, streamflow normally becomes unreliable and insufficient for sustaining people's quality of life, the prevention of disease and other human health problems, the maintenance of irrigation works, and meeting urban and industrial needs. Needed groundwater levels also decline, resulting in the failure of springs, wells and water development for livestock.

Severe watershed degradation has taken place on large portions of the East African highlands because of the loss of protective vegetative cover in the region. At one time, nearly 75% of Ethiopia was covered with dense forest vegetation that helped to mitigate excessive soil erosion on steep landscapes. Unfortunately, from the 1950s to 1989, this cover has diminished from 16% to 2.7% of Ethiopia's land area for various reasons (Ethiopia Forestry Action Program, 1993). As a consequence of this loss of vegetation, soil loss from the Amhara Plateau has resulted in increased silt levels that the Nile River has carried to fertilize the agricultural flood plain of Egypt for millennia. Fortunately, the Aswan High Dam in Egypt has helped to control the Nile's floods and at the same time trap the silt. Its construction created a reservoir of more than 5000 km^2 that traps an estimated 90 million tonnes of sediment each year, giving the reservoir a life expectancy of 500 years in a land with a cultural history of more than 5000 years (Baecher et al., 2000). However, this rate of sedimentation is likely to increase in the future in the absence of improved stewardship of the watershed landscapes.

Soil loss and nutrient depletion in humid tropical regions

Large numbers of people live on forested slopes in the humid tropical regions of the world. These people clear the slopes for fuel, fodder and small-scale agricultural cultivation. Fire is frequently allowed to escape from the agricultural fields and burn indiscriminately, forests are grazed by livestock to levels that prevent their renewal, and roads and trails are often established with little concern for the soil loss and nutrient depletion that they create. Removing the living biomass on these slopes in much of the humid tropics means rapid soil loss and nutrient depletion, since some 80–90% of the nutrients are found in the living biomass. If they are not returned to the soil to decay, these nutrients are lost. Consequences of these actions include accelerated soil loss and land deterioration, environmental degradation and further impoverishment of the rural inhabitants themselves. The available nutrients tied up in vegetative canopies are returned to the soil only after the slash residues from timber harvesting are burned. As a result of this turnover, sustained shifting cultivation has commonly been practised in humid tropical regions for thousands of years. However, this form of agriculture becomes a *serious environmental threat* when human population pressure on the land becomes too great to allow a sufficiently long fallow period between slash-and-burn cycles. There is evidence that such pressures contributed to the collapse of several civilizations – notably the Mayan civilization of Central America and the ancient Khmer Empire of Cambodia – whose agricultural practices led to cementation and loss of fertility of the soils that they farmed.

Problems of sustaining people's livelihood

Between 50% and 80% of the rural human populations of the world live on lands either too steep or too dry or on lands where the soil is too poor to support more than the minimal level of existence. The more children these communities have and the harder they work to support their families, the poorer they become. The largely fragile environments in which they often live are being subjected to increasing mistreatment that leads to a vicious downward spiral in their well-being. We see the examples of this all over the world. In some cases, entire regions are devastated and moving in the downward spiral, while it is affecting pockets of poor communities in otherwise progressing regions elsewhere. Eventually, some of these desperately poor, increasing human populations are forced to migrate to lands that are even more *marginal* in terms of agricultural productivity.

The situation in western India is one example of this challenge. The arid regions of India that include the Thar Desert of western Rajasthan have population densities of more than 60 people per square kilometre. The practical consequence of the pressure that this population exerts on the land has been the extension of farming activities onto largely *submarginal* lands, which are often suitable only for marginal livestock production, making this perhaps the dustiest area in the world. Meanwhile, the number of grazing livestock frequently swells as the land available for sustained forage shrinks. The area in western Rajasthan that is available exclusively for livestock grazing has steadily dropped since the 1950s while the human population continues to grow (Chandra and Bhatia, 2000). The farmed area of the region has expanded by about 15% at the same time, shrinking the area for livestock grazing even more. The livelihood of tens of millions living on this arid region is likely to remain at the dismal level presently faced as long as these current land-use patterns continue to persist. At worst, a prolonged drought – which is bound to occur – will mercilessly rebalance the number of people with the available resources. As it is, relief programmes for these lands and people are seriously draining the government's funds and available food stores.

Other challenges

There are many other challenges which will be elaborated as the chapters unfold. However, the above examples suffice to make the point that serious environmental problems are created by people mistreating and misusing their land and water resources. These can lead to serious social and economic problems and, ultimately, at times to starvation and death among the poor of the world.

Meeting the Challenges: Some Examples

We know technically how to address and overcome most of the biophysical challenges confronted and we have good insight on what works and does not work institutionally when people adopt concerted, cooperative actions in an IWM framework (see Chapter 2). However, we know less about the associated socio-economic challenges that are often the causes of many biophysical challenges. Evidence indicates that well-planned, politically supported and effectively implemented land and water management activities within an IWM context can help to reduce or eliminate some of the key problems. Such approaches to land and water stewardship are the subject of this book.

The concepts of IWM furnish a framework for attaining sustainable development and conservation of natural resources, while watershed management practices, projects and programmes provide the tools for making the framework operational. Institutional mechanisms including regulations, market and non-market incentives, and public investment and management provide the institutional context for formulating and implementing the needed activities of different groups of people (see Chapters 3 and 4). As emphasized earlier, implementation of successful IWM depends on knowledge both of the physical and biological characteristics of watersheds and of the institutional factors, such as the nature of governments and their governance quality and the cultural backgrounds and economic situations of local populations. These must be fully integrated into viable solutions that meet the environmental, economic and social objectives of the people. How these factors are interrelated can be illustrated by looking at a few examples from different parts of the world – some of which have been more successful than others.

Alleviating water scarcity in arid regions

Water relationships in the arid regions of the world are perhaps more critical to a greater number of people on earth than those of more humid regions. Water is usually in short supply and in great demand by people and their livestock in these fragile regions. People living in arid regions do not often think in terms of the IWM approach to achieving better land stewardship in attempting to alleviate these problems. However, it is here where the flows of water and other natural resources to people are most frequently limited in both quality and quantity. Furthermore, the need to coordinate land, water and other natural resource management efforts is important both economically and politically to the inhabitants of these countries.

Nowhere is this need as crucial as it is in the Middle East. Syria, Jordan, Israel and the neighbouring countries of Turkey and Iraq are faced with a situation in which most of their surface and groundwater resources have been fully allocated (Bureau for the Near East, 1993; Metz, 2000). Nevertheless, the populations of people and their livestock continue to increase in this region. It follows that almost any type of land-use change must account fully for the accompanying changes in water supplies. Opportunities for necessary economic development require coordination among the countries and that environmental effects become part of the planning process. Prospects for conflicts over water in this critical region are not trivial. The politics of water is paramount in the 21st century in the Middle East.

Providing necessary natural resources to people

Thailand, Indonesia, the Philippines and other countries in South-east Asia have extensive watersheds that are a key to providing necessary natural resources to their people. High-quality water flowing from these watersheds is a major resource along with wood and other forest products, agricultural crops and wildlife habitats. Realizing the importance of these natural resources and the need to sustain flows of high-quality water, government agencies and non-governmental organizations (NGOs) in Thailand implemented a series of watershed management practices, projects and programmes on a countrywide basis in the early 1980s to protect crucial soil resources while sustaining water for downstream users, raw material for wood products and other natural resources in demand by the people. These efforts began at a relatively small scale on comparatively small areas of high priority relative to their potential to deliver these benefits, and then progressed to larger scales and larger areas on the basis of incorporating the initial results and findings into the planning process.

Reconciling timber production and other watershed values

Timber production in the mountain ranges of Oregon in western USA has been a main segment of the regional economy. However, some harvesting activities necessary to obtain this timber threaten another main segment of Oregon's economy, the multi-million-dollar salmon fishery (Meehan, 1991; Stouder *et al.*, 1997; Beschta, 2000). Clear-cutting of trees on steep slopes with the accompanying road construction has led to many landslides that deposit soil and other debris into stream channels and rivers. Although landslides are natural occurrences on this terrain, timber harvesting and road construction can accelerate their frequency of occurrence and extent. When this happens, landslides become a watershed management problem.

What measures are needed to reduce landslides, road-related soil erosion and sedimentation to protect the salmon fishery and other resource use, and to sustain timber production in the western USA has become a relevant question to be answered by the people in this region. In meeting this challenge, they have considered the physical and biological input–output relationships involved and then attempted to find ways of altering these relationships by changing their land-use practices or modifying their existing ways of doing things. For example, forest practices have been adopted that reduce road-related erosion and delivery of sediment, and that promote stream crossing that allows fish passage (Ice *et al.*, 2004). Similar best management practices (BMPs) have been adopted in the USA for maintaining streamside buffers that also protect fish habitat. Importantly, the appropriate measures have been tempered by the social and economic implications of the alternatives available and the political realities of the situation. Some measures have been more acceptable than others by the affected groups of people, while other groups of people have accepted different measures. Recognized trade-offs have become a prerequisite to reaching a consensus on appropriate solutions. The costs of different alternatives have also varied, so the cost-effectiveness of the alternatives is considered in relation to meeting budget constraints and political criteria.

Sustaining agricultural production on marginal lands

There are continuing challenges to achieving effective management on the watersheds in the Amazon Basin of South America. By almost any account, the soils found in most of the Amazon Basin are low in fertility, with most of the soils with medium- to high-fertility found on the narrow plains along the banks of rivers. Unfortunately, the use of these soils for large-scale agriculture requires large financial expenditures for the necessary drainage and continuing flood control. Nevertheless, agricultural development programmes have been tried to encourage farmers from other regions of Brazil to migrate to selected watersheds in the Amazon Basin. Since the early 1970s, 50,000 families have settled along a proposed highway between Peru and the Atlantic Ocean

(Brooks *et al.*, 2003). With limited financial and administrative resources and less technical knowledge of tropical farming techniques, the most successful of the programmes only attain agricultural production at subsistence levels. It should not be surprising, therefore, that increasingly more people find it difficult if not impossible to make a living and, as a consequence, abandon their plots after inappropriate agricultural cultivation has degraded the soil.

The loss of forests in the mountain zone of Sumatra, Indonesia, and the consequential loss of soils and diminished water quality are being address through improved agroforestry practices and corresponding changes in policy (Agus and van Noordwijk, 2005). Conflicts between farmers who cleared the forests to grow annual crops and state forestry officials in Sumberjaya, West Lampung Province, led to large-scale evictions of farmers from the land. Following political reforms, farmers returned to the area. A project ensued in which agroforestry practices were introduced as alternatives to forest clearing and were designed to achieve sustainable production and to restore impaired watershed functions. A process of negotiation between stakeholders led to a combined research and monitoring project aimed at improving land production through practices that met farmer's preferences while reducing soil erosion. With the promise of improved land tenure, farmers have adopted new practices that provide alternatives to forest clearing or farming that led to land and water degradation in the past. Monitoring of land-use change, streamflow quantity and water quality is ongoing.

A success story

Watershed management interventions involving the rehabilitation of degraded landscapes to restore flows of natural resources and, in doing so, lead to better stewardship of the watersheds have been successful in some countries. The Republic of Korea is an example of one country that has made progress in sound land and water stewardship of its mountainous watersheds within the framework of the IWM approach. The urgency for achieving this purpose was related to both the need to protect an expanding network of major hydropower installations in the country and the relatively small but highly productive areas of irrigated agricultural lowlands that require water from the high mountains in the north (Gregersen, 1982; Bochet, 1983). The Korean approach was one of integrating watershed management practices with environmentally sound forest management activities and effective community-development efforts through its comprehensive community movement programme. This holistic programme, called the *Saemaul Undong* or New Community Movement, was successful for a number of reasons. An exhaustive case study summed up the main reasons as follows (Gregersen, 1982):

- Broad-based approach (through Saemaul Undong). Forestry activities were tied to a number of other dimensions of watershed management and community development, and villages (e.g. through local women's associations) were involved in seedling production etc.
- Incremental approach (*realistic potentials* approach). At each village level, the initial activity was tailored to expectations for success to avoid failures that could discourage or distract villagers.
- Top-down combined with bottom-up approach. The programme involved government officials from highest to lowest levels. This was paralleled by a

bottom-up approach using the hierarchy of village forestry associations (VFAs) up to the National Federation of VFAs (NFVFAs). Among other things, the NFVFA provided the mechanism for bargaining to get local VFAs the best prices possible for their outputs.

- Cooperation between government and private citizens and NGOs. (See previous item.) There was a concerted effort to involve both government and NGOs to ensure joint ownership of *success*. Linkages between government and the VFA structure were established at all levels.
- Emphasis on short-term gains. It was recognized that the incentive for participation by locals depended on some gains derived early in the programme (before the benefits from tree planting). Sale of shiitake mushrooms, forest stones and Kudzu fibre were part of the programme and provided early returns to the villages that participated.
- Appropriate technology. Careful attention was given to getting the technologies right in terms of reducing uncertainty and ensuring good returns from the programme. For example, species were chosen for fuelwood plantations so that high levels of survival after planting could be obtained and could have multiple uses in case all trees were not needed for fuelwood or charcoal.
- Adequate technical assistance and extension services. Effort was spent to transfer the appropriate technologies to the local communities in a way that would minimize chances of failure.
- Thorough logistical planning. Proper planning was implemented in terms of delivery of seedlings and providing training early enough.
- Appropriate subsidies (and required reinvestment of gains). Villages received subsidized support for the fuelwood programme only if they could show local investment in community improvement. In addition, VFAs were required to reinvest part of the proceeds from their various activities back into community improvement programmes (schools etc.).

We come back to these factors of success, since it turns out that they characterize many if not most of the successful large-scale IWM projects.

2 Land Use, Watershed Management and Cumulative Effects

The primary goal of IWM is providing people with goods and services in ways that maintain the long-term productive capability of the natural resources. People want a great many goods and services from a watershed, including water for consumption, irrigation and hydropower; raw materials for wood products; agricultural crops and meat and dairy products; recreation and tourism; and maintenance of biodiversity. Depending on the situation, therefore, the objectives of IWM are likely to be one or more of the following:

- Developing, managing and sustaining production systems that are well suited to the existing environment and resource base and that can be sustained for future generations.
- Sustaining or increasing supplies of high-quality water and improving the quality of water in streams, lakes and other water bodies.
- Preventing excessive soil erosion to protect the productive potentials of the land and reduce downstream sedimentation.
- Rehabilitating watershed lands, stream channels, wetlands and riparian systems that have become degraded.

Many of these objectives are also associated with soil and water conservation practices aimed at local farmers, livestock producers, foresters, engineers or others who exploit natural resources on a watershed for different purposes. Indeed, soil and water conservation represents an integral part of watershed management. However, what distinguishes the IWM approach from other land management practices is its holistic consideration of the linkage among all of the activities within a watershed and how these linkages affect each other and the sustainability of the natural resource base on which people depend. Achieving the goals of IWM requires planning for the effective and efficient use of the land, water and other natural resources, coordinating the management activities and benefits among different groups of people, dealing with many land uses, and, quite often, working within more than one political jurisdiction.

Interactions of Land Uses

One can think of land uses on a watershed in many different ways. In the following discussion, however, we think of them in terms of interconnections between the uses of land, water and other natural resources within the boundaries of a watershed. The watershed, therefore, can be thought of as a collection of land-use units of varying physical sizes, each of which is defined by its ownership or *control* and use(s) of the land on the unit. Relationships between a land-use unit and the rest of the watershed are conceptually illustrated in Fig. 2.1. Note that any upstream activity on other land-use units influences the

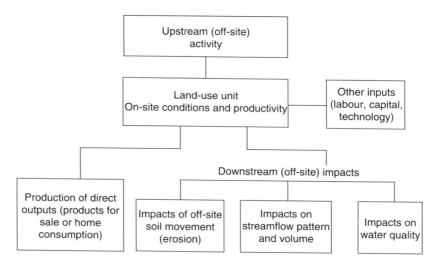

Fig. 2.1. Interactions among land-use units on a watershed.

condition and productivity of the downslope land-use unit in this figure. Among those influences are the effects on the rates of soil erosion that (in turn) deposits materials on the downstream land-use unit and, as a consequence, changes the soil quality, water flow patterns and water availability on the watershed. All of these and other effects can impact on people's food availability, health standards and income. Also note that land, water and other natural resources management practices implemented on the downslope land-use unit in Fig. 2.1 impact on what happens to productivity and sustainability of the natural resources on site whether it is related to agricultural, livestock grazing, forestry or other uses. Finally note that what is done by way of management practices on this land-use unit can influence soil erosion and sediment movements, water flows and water quality characteristics. While what happens on any land-use unit might in itself have a small impact on downstream land-use units, these impacts are often significant when aggregated; hence the significance given in this chapter to *cumulative effects*.

We look at the on-site and off-site impacts of land, water and natural resources management on the land-use units within a watershed in the following sections. These management practices are the core of the act locally part of IWM (see Chapter 1). People carry out these management practices to obtain the outputs needed for their survival, well-being and incomes. Therefore, the land-use units can be cultivated for the production of agricultural crops, be a source of forage for grazing livestock or be a source of raw materials for wood products. A key to obtaining all or combinations of these outputs is the availability of high-quality water, a topic that is discussed first.

Availability of high-quality water

Water limits much of what people can do. Therefore, sustainability of high-quality water is critical to the welfare and, ultimately, to the survival of people. Too much water at any given time (e.g. floods) can be a disaster; too little water available at a given time can also be a disaster. The keys to good IWM are moderating flows of water, maintaining the quality of water, preventing groundwater drawdown and pollution, and mitigating water flows to minimize soil erosion. Watershed management plays a critical role in managing, developing and/or increasing supplies of the high-quality water that are necessary for improving food security and other welfare benefits for people living both on upland watershed and downstream areas. Poverty alleviation in poor countries depends very much on proper management of water and associated resources; and this includes management of water resources in an equitable way that does not hurt the poor. Increasing demands for water have often resulted in the two kinds of primary responses as illustrated in Fig. 2.2. Unfortunately, conservation of water resources by reducing per

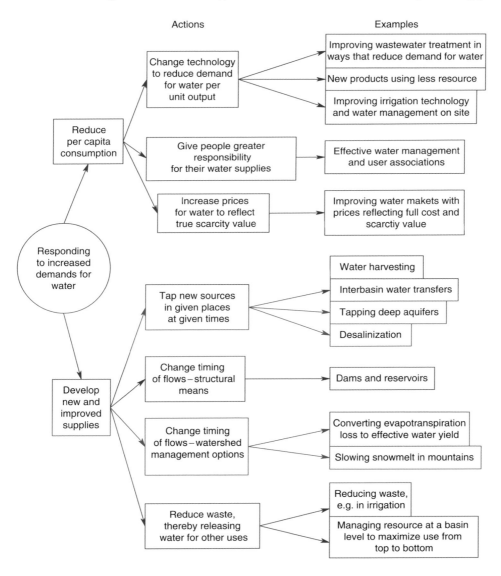

Fig. 2.2. Responses to the increased demands for water. (Adapted from Gregersen *et al.*, 2000.)

capita consumption – one primary response – has had limited utility in meeting people's demands in many parts of the world. However, numerous methods have been used in developing water supplies in the absence of locally dependable perennial streams or abundant groundwater resources – the other primary response.

We must also consider the issues that are associated with patterns of water flow in terms of both high and low water flows. Coping with high flows and, in the extreme, flooding has brought about an array of structural and non-structural measures for its control as shown in Fig. 2.2. One structural response is controlling the impacts of flooding with dams and reservoirs to store floodwater and then release the water in periods of low flows. Whatever the stated purpose of a dam and reservoir system, reservoir operations need to take into account their potential effects on downstream water quality and aquatic ecosystems. Sustaining minimum *in-stream flows* as specified by regulations is often an objective of reservoir management. A non-structural approach to controlling the impacts of flooding is through effective flood plain management in which the focus is placed on managing people's behaviour to minimize flood hazards by limiting development in flood-prone areas. Many cases exist currently where governments are quietly buying up critical lands in flood plains.

Water quality is still another consideration in sustaining water for people's use and for protecting important aquatic ecosystems. The impairment of water quality impacts on the utility of water supplies – depending on its anticipated use – and the health of people. More than 3 billion people worldwide are denied access to clean water, with the most acute problems occurring in *developing countries* where 90% of wastewater from industrial and/or human use is discharged into streams without treatment (Johnson *et al.*, 2001). Of the more than 3 million annual deaths attributed to polluted water supplies and poor sanitation, in excess of 2 million are children (van Damme, 2001). Efforts are therefore underway globally to attract the political support necessary to develop the necessary infrastructure and management to improve water quality for human consumption and ecosystem services. In fact, one of the eight Millennium Development Goals relates directly to water quality and available quantity.

To achieve the goals of sustaining or improving water quality requires efforts to reduce both point sources and non-point sources of pollution. Point sources of pollution such as the discharge of wastewater into streams or rivers can often be controlled through regulatory actions. However, non-point sources of pollution are more difficult to address as they are the result of multiple factors including, importantly, the land-use activities that occur within the boundaries of a watershed. Meeting specified water-quality standards (Box 2.1) through the control of non-point sources of pollution requires that people have the knowledge and resources to make the necessary changes in land-use practices to achieve this purpose. Water-quality standards have been established in many countries to monitor water quality and take remedial actions when necessary.

While people's land-use practices can affect the availability of high-quality water, people's efforts to develop and/or increase water supplies for their use affect the types of land use that can take place on a watershed. The construction of dams and canals can promote dramatic changes in land use, for example, through the expansion of irrigation systems and urban development. The drainage of excess water from water-logged soils on a watershed can cause changes in land use, converting wetlands into agricultural croplands or pastures in some cases. As we see more intensive agricultural production and other development activities on a watershed, water resources

Box 2.1. Water quality standards: one example.

The general purpose of water quality standards is to protect, maintain and, when possible, improve the quality of surface waters. The Department of the Environment for the State of Maryland has established the water quality standards for the state – which lies along the coast of the Atlantic Ocean – that include the three components of designed uses, water quality criteria to protect the designated uses, and anti-degradation policy.

Among the designated uses of water flowing from upstream watersheds in Maryland are:

● Water contact re-creation and protection of non-tidal warm water aquatic life;
● Water contact re-creation, protection of aquatic life and public water supply;
● Tidal freshwater estuary;
● Non-tidal cold water;
● Non-tidal cold water and public water supply;
● Re-creational trout waters;
● Re-creational trout waters and public water supply.

Numeric water quality criteria set the minimum water quality to meet these designated uses. Maryland has several numerical criteria for the protection of aquatic life and human health. Water quality criteria are published online for toxic constituents, dissolved oxygen, turbidity, bacteria and temperature. *Narrative criteria* apply where specific numeric criteria are not available such as oil, grease and odour. Changes in water quality standards and water quality criteria are implemented through regulatory changes that are subject to the state's promulgation process.

Maryland's antidegradation policy assures that water quality continues to support the designated uses. Three tiers of protection are considered:

● Tier 1 specifies the minimum standard that must be met in support of *balanced* indigenous populations; this support is often referred to as *fishable–swimmable*.
● Tier 2 protects water that is better than the minimum specified for that designated use.
● Tier 3 is currently in the developmental process, and is called *outstanding national resource water*.

Fact sheets on water quality including answers to frequently asked questions are provided to people on request. Public information meetings regarding proposed changes in watershed quality standards and water quality criteria are scheduled when needed to obtain comments on the proposed changes.

projects become increasingly important to the economy of an area. Figure 2.3 illustrates the economic implications of the linkages between land-use practices to sustain or improve productivity and water development or use. It is at this point that we often see political interest in these land-use activities that can adversely affect projects, which (in turn) focuses further attention on the need for IWM practices that recognize the holistic nature of a watershed and what takes place on it. Interactions among land use and water development have also become important considerations of watershed management at large scales on river basins. We consider the role and consequences of other land uses at the *local* level as well as larger watersheds and river basins in the following sections.

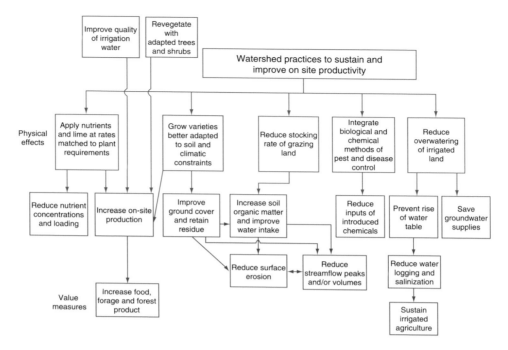

Fig. 2.3. Land and water management practices to sustain and improve on-site productivity.

Agricultural cropping

Rain-fed and dispersed agricultural cropping is a common land use on many upstream watersheds. While small in their individual contributions to the agricultural economy of a region, the aggregate production of all agricultural cropping on upland watersheds can be comparatively large (Altieri, 1988; Spedding, 1988; Ffolliott *et al.*, 1995a). Either *subsistence* or *commercial farming* is practised by these farmers, depending largely on the capacity of the land to produce agricultural crops and the level of capital available to the farmers to do so. The crops produced are consumed at home to meet the immediate needs of subsistence farmers and/or the occasional surplus might be sold at a marketplace. Large-scale commercial farming is less common on upland watersheds because of the frequent need for large irrigation facilities, diversified marketplaces and the necessary infrastructure for the transportation of managerial inputs and production outputs.

Agricultural cropping systems used by a farmer are dictated by the local climatological conditions, inherent soil capabilities, as well as the farmer's needs, abilities and perceptions of agricultural practices. While the agricultural cropping systems applied on upstream watersheds are endless in their strategies and methods of implementation, they can be grouped into categories of *sedentary agriculture* and *shifting cultivation* for discussion purposes. Sedentary agriculture is practised when the soil fertility and precipitation and temperature regimes allow crops to be grown in place on a more or less continuous basis and people settle, building more or less permanent homes and communities. One crop a year is usually grown when rainfall is sufficient, with multiple cropping often possible in more humid regions.

On the other hand, shifting cultivation requires the farmers to move from one site on a watershed to another once the capacity of the soil to produce the needed agricultural crops at subsistence levels on the original site is lost – often 3–5 years in the tropics. Cycles of shifting cultivation include clearing of trees on the site, a burning of the remaining vegetation with the ash serving as fertilizer, and planting of the agricultural crops. When soil fertility declines to the point of severely limiting crop production, the farmer moves on to repeat the cycle elsewhere, most of the time returning to the original piece of land once a reasonable fallow period has passed and the nutrients have built up in the surface soils again.

Farmers engage in agricultural cropping on watersheds that might also be used to grow forage or fodder for livestock production, trees for commercial purposes or environmental protection. Potential conflicts can be encountered when this is the situation. However, small-scale agriculture can be compatible with other watershed management activities when it is practised only on the sites most suitable for agricultural cropping. One way by which sustained agricultural cropping has been achieved on a watershed basis is through a geographic separation of agriculture from other land uses with the other watershed strata put to the use(s) to which they are most suited. Another option is alternating or rotating agricultural cropping with other land uses of a watershed being managed to maintain water flow, livestock production, forestry activities, etc. This option can, however, evolve into a shifting cropping system. A third option of achieving combined production involves agroforestry systems that can consist of different combinations of agricultural cropping, livestock production, forestry activities and other land practices concurrently or over time.

Livestock production

Maintaining livestock is a way of life for many rural societies and, therefore, proper livestock-grazing practices are necessary to sustaining livestock benefits such as the production of milk, meat or wool on a watershed landscape. Either dispersed livestock grazing on open rangelands or confined livestock grazing in small pastures is normally practised. Some watersheds are able to sustain varying combinations of domestic livestock and/or indigenous (wild) herbivores with appropriate grazing management. For example, while possibly complicating management procedures, the inclusion of sheep or goats with cattle can increase total livestock production without adversely impacting the availability of water resources, forage or fodder, or the watershed condition (Pratt and Gwynne, 1977; Jacobs, 1986; McLeod, 1990; Holechek *et al.*, 1998). A better distribution of animals on the watershed can often be achieved by doing so, resulting in a more uniform use of water and forage resources.

Degradation of a watershed to a poor condition – when this occurs – is most commonly the result of overstocking livestock, which can cause an eventual loss of high-valued forage species and, as a consequence, a lowering of livestock growth and development. Additionally, trampling by an excess number of livestock compacts soil surfaces, reduces infiltration of surface water into the soil and increases overland flows of water and soil losses from the trampled site (Branson *et al.*, 1981; Jacobs, 1986; Holechek *et al.*, 1998). Poor choices of watering areas for livestock and improper management of areas adjacent to water can cause further damage.

Attaining sustainable livestock production on a watershed requires the following:

- The area to be grazed be stocked only with the number of livestock that can be supported on a sustainable basis.
- Livestock grazing be permitted only when adequate forage resources are available.
- Livestock be properly distributed on the watershed and not be allowed to concentrate along streambanks or other watering sites where they can cause soil erosion, sedimentation or other pollution.

Wood production and other forestry activities

Trees are a source of construction materials, poles and posts, and fuel for the people living both on upland watersheds and downstream. They are also a source of fodder for livestock and browse for wildlife populations when herbaceous forage is not available. Fruits, leaves, young shoots and roots of trees are often valuable food reserves for people. Additionally, trees play important roles in maintaining the delicate ecological balance of many watershed environments. The roots of trees hold soil in place and, as a result, control soil erosion and help to stabilize steep slopes. Trees in windbreak (shelter belt) plantings protect the site from accelerated aeolian erosion, lessen evapotranspiration (ET) rates and moderate air temperature extremes.

It is unfortunate that trees have been *mined* more than *managed* as a renewable natural resource on many watersheds in the world; they have been too heavily harvested by people without concern for their renewal. The conversion of forests and woodlands to agricultural croplands and livestock-grazing lands on sites that are not able to sustain either agricultural or livestock production can cause a degradation of soil and water resources. Incidences of wildfire or fire that is purposely set by people to convert forests or woodlands to agricultural croplands or livestock pastures can also damage soil structures and increase soil erosion from the burned sites. In many parts of the world, the encroachment of urban areas in forest lands can dramatically increase runoff and soil erosion and degrade water quality.

Environmentally sound forestry practices to sustain wood production and other forestry activities on a watershed require appropriate managerial knowledge of the tree species comprising the forest or woodland. Knowledge of reproductive methods, silvicultural cuttings and appropriate applications of other cultural treatments is available to professional foresters and, at times, the people who are users of the land on many watersheds throughout the world. Protection measures against wildfire, disease and insect infestations are also known in many instances. However, problems occur when this knowledge is not available, it is incomplete in its content or it must be extrapolated form elsewhere and applied without testing and validation. Wood production and other benefits from trees are likely to become unsustainable in these situations.

Agroforestry practices

Agricultural cropping, livestock grazing and/or forestry activities often occur in varying combinations within a watershed boundary. Watersheds, therefore, are often mosaics of these and other forms of land use. Some of the best opportunities for people living on watersheds to match their land uses with the capacities of a watershed to achieve needed productivity levels and benefits – and to benefit people living downstream because of reduced runoff, sediment accumulations and other cumulative effects at the same time – often involves the integration of agroforestry practices into management of a watershed (Nair, 1989; MacDicken and Vergara, 1990; Gordon and Newman, 1997). *Agroforestry* is a system of land-use practices where trees or other woody perennials are grown on the same piece of land as agricultural crops, livestock – or a combination thereof – either simultaneously or sequentially (Box 2.2). As such, agroforestry systems and practices become effective and efficient combined production systems that have a bearing on sustaining the welfare of watershed inhabitants.

The so-called *multi-purpose tree species* are often grown to benefit the people on the watershed. These trees themselves are a form of agroforestry in that they can be useful for wood production and protecting and often improving the fertility of soil while providing leaves and small branches for fodder without impairing agricultural cropping (MacDicken and Vergara, 1990; Ffolliott *et al.*, 1995a). Many multi-purpose trees fix nitrogen in the soil in addition to providing benefits to local people, a further contribution to sustaining good watershed management.

Urban development and roads

Watersheds or portions of watersheds can be converted into roads and other transportation corridors, become sites where housing and other structures are built and/or be dammed to create reservoirs for water storage. All of these infrastructures, as components of urban development, can have important on-site and off-site impacts.

Box 2.2. Agroforestry systems on upstream watersheds: common examples. (From Nair, 1989; MacDicken and Vergara, 1990; Gordon and Newman, 1997.)

Agroforestry systems on upstream watershed lands are mostly *agrisilvicultural*, *silvopastoral* and *agrosilvipastoral* in their structure. Agrisilvicultural practices are combinations of agricultural crops and forestry activities, with the agricultural crops dominating. Silvopastoral combinations are forestry activities and livestock production, with a dominate land-use of forestry. Agrosilvipastoral practices within these agroforestry systems include agricultural cropping, forestry and livestock production in varying combinations of dominance. Arrangements of the components of agroforestry differ in space (random, alternate rows or border-tree planting) and time (coincident, concomitant or sequential). Agroforestry practices also differ in their function; that is, they can function to satisfy one or more of the needs of people, ameliorate microclimates, retain soil and water resources, or combinations of these and other protective functions. From a socio-economic standpoint, agroforestry practices are *subsistent*, *commercial* or *intermediate*, depending on whether the outputs obtained meet the basic needs of the people, are made available for sale at a marketplace or a combination of the two.

For example, while roads facilitate the necessary movement of managerial inputs and production outputs, they often disrupt the hydrologic functioning of a watershed and cause severe soil erosion and stream-channel degradation if they are located on steep slopes, on erodible soils, in narrow riparian corridors, landslides on steep terrain, or at stream crossings. These latter impacts often exceed those caused by all of the other land-use activities on a watershed combined (Satterlund and Adams, 2000; Bell, 2000; Brooks *et al.*, 2003). Up to 90% of the sediment that is produced on a watershed can originate from roads at stream crossings (Egan, 1999; Grace, 2000; Chang, 2003). Roads that facilitate increased accessibility to the more remote parts of a watershed can foster more rapid and often ill-planned human settlements and the accompanying increased exploitation of natural resources. It is imperative, therefore, that roads and other transportation corridors, housing and other structures, and dams and reservoirs are carefully planned, properly constructed and continuously maintained to reduce these adverse impacts.

Linkages Between Land Use, Soil and Water

People's use of land, water and other natural resources can sustain production of food and other desired outputs if their use is carried out in a manner that is not detrimental to these resources and the environment. The linkages between land use, soil, the flow of water from a watershed and, ultimately, the quantity and quality of water and other natural resources must be recognized and appreciated by both the people using the land and the people planning and managing the proper use of the natural resources on a watershed. Transforming this knowledge into sustainable and environmentally sound management of land, water and other natural resources use that is guided by IWM principles enables people to grow agricultural crops, graze livestock, manage forests and maintain, develop and/or increase water resources in a sustainable manner.

Achieving watershed management objectives

Watershed management practices serve as the *tools* for achieving environmentally sound land-use and watershed management goals and objectives. These management practices include a variety of non-structural (land use and vegetative) and structural (engineering) measures undertaken to meet these goals and objectives (Table 2.1). These measures are implemented either singularly or in varying combinations to achieve the objectives stated in the introduction of this chapter.

Two questions that might be asked at this point are: 'who are watershed managers?' and 'what are their responsibilities in achieving watershed management objectives?' Ultimately, watershed management resides in the hands of the users of the land, i.e. the farmers, livestock producers, foresters, urban developers, etc. We might infrequently see people with the title of *watershed manager* on the land, although all of the users of land, water and natural resources should recognize that they also play a role of watershed manager as we intend to illustrate in this book. There are exceptions, however, when a technically trained person has specific responsibilities to manage a watershed for a stated purpose, for example, someone managing municipal

Table 2.1. Watershed management objectives and non-structural and structural practices. (Adapted from Brooks *et al.*, 1990.)

Objectives	Measures	
	Vegetative/land use management	Structural
Maintain or increase land productivity	• Agroforestry practices (e.g. alley-cropping, agrosilvopastoral systems) • Reforestation or afforestation to meet fuel, fodder and fiber needs • Soil conservation practices (e.g. strip cropping, no-till or minimum tillage cropping, mulching or cover crops, vegetation to stabilize structural conservation measures) • Limiting grazing to sustainable levels	• Terraces (bench, broad-based) • Contour ditches and furrows • Gully-control structures and grassed waterways
Assure adequate quantities of usable water	• Encouraging low water-consuming species • Using appropriate land-use measures to protect reservoirs and channels	• Water harvesting, spreading and irrigation measures • Water-harvesting systems • Reservoir and water diversion structures • Irrigation facilities • Wells • Encouraging water-saving technologies
Assure adequate water quality	• Maintaining or establishing vegetative cover in key areas (e.g. streambanks) • Controlling waste (human, livestock, mining, etc.) disposal • Using natural forests and wetlands as secondary treatment systems of wastewater • Controlling grazing and developing guidelines for riparian systems	• Water treatment facilities • Developing alternate supplies (e.g. wells, water catchments)
Reduce flooding and flood damage	• Revegetating or maintaining vegetative cover to enhance infiltration and water consumption by plants • Zoning/regulating flood plain use • Protecting and maintaining wetlands	• Reservoir flood control storage • Water diversion structures • Levees • Gully-control structures • Improving channels (channelization)
Reduce the incidence of landslides	• Reforestation or afforestation for soil stabilization	• Bench terraces

Continued

Table 2.1. *Continued*

Objectives	Measures	
	Vegetative/land use management	Structural
Reduce downstream sediment delivery	• Maintaining good vegetative cover to promote infiltration of rainfall • Restricting residence and productive activities on steep unstable slopes • Maintain vegetative cover on hillslopes; utilize similar practices to maintain or increase land productivity • Maintain healthy riparian vegetative systems and maintain perennial cover on flood plains	• Grassed waterways, drop structures, etc. to control overland flow • Use similar structures as above to maintain or increase land productivity • Channel restoration

watersheds to provide drinking water to people in urban areas. In any case, there are many roles for the people in practising watershed management from farmers to livestock grazers, forest managers, urban developers and, importantly, government and NGOs with the responsibility for achieving the following objectives.

Maintaining good watershed condition

Maintaining a watershed in good condition is a necessity if we wish to sustain the flows of high-quality water, agricultural cropping, livestock grazing, wood production and other beneficial purposes in an environmentally sound manner on the watershed. *Watershed condition* is a relative term that indicates the health (status) of a watershed in terms of its hydrologic function and soil productivity. *Hydrologic function* relates to the watershed's ability to receive and process precipitation into streamflow and oftentimes groundwater. *Soil productivity* reflects the capabilities of a watershed for sustaining plant growth, plant communities and the natural sequences of plant communities. People must recognize that the above terms should be considered in the context of the environmental setting of the watershed and that each has inherent limits that are set by climatic, topographic, geologic and other conditions. A watershed in good condition is one in which:

- Precipitation infiltrates at the maximum rate for soils and settings in the watershed, thereby minimizing surface runoff.
- Precipitation does not contribute excessively to soil erosion; maintaining good vegetative cover on the soil surface protects the soil from direct rainfall splash and there is minimal surface runoff to dislodge and move soil particles.
- Streamflow response to precipitation input is relatively slow; watershed conditions are such that the watershed maintains its hydrologic stability.
- A sustained level of minimum flows (usually from groundwater) in perennial streams is maintained between storms; low flows are sustained according to natural climatic variations.
- Sediment and nutrient loading to streams is minimal due to watershed conditions and a healthy riparian system.

It is clear that a watershed in arid regions with steep slopes, naturally sparse vegetative cover and shallow soils needs to be judged differently in its condition than a watershed in more humid regions with forest cover and deep soils. However, watersheds in similar settings would be considered in poor condition if they deviated significantly from their natural or undisturbed characteristics. Watershed management practices that sustain a watershed in good condition are those that:

- Maintain a protective vegetative cover on headwater tributaries, on hillslopes and, in general, across the watershed landscape; improper agricultural cultivation and excessive timber harvesting and livestock overgrazing are avoided;
- Minimize soil loss on hillslopes associated with mining activities; discharges of excessive sediment and heavy metals into stream channels or other bodies of water are not allowed;
- Control erosion through engineering measures such as the construction of check dams as the first step to control erosion on sites experiencing excessive soil loss; engineering measures are followed by more permanent vegetative measures such as seeding or planting of vegetative covers;
- Plan and construct roads and other transportation corridors to minimize cuts and fills on hillslopes, minimize stream crossings and avoid riparian areas;
- Minimize intensive re-creational use that compacts soils and causes excessive transport of nutrients to streams, lakes and other water bodies and concentrates people in fragile riparian corridors.

These characteristics are consistent with the primary objective of IWM: to sustain flows of high-quality water and other natural resources from watershed landscapes. They must be appropriately incorporated into the planning process (see Chapter 4) to ensure the proper management of the watersheds.

Sustaining and improving on-site productivity

Vegetative and engineering measures are available to sustain and/or increase the on-site productivity of watersheds in terms of small-scale agricultural cropping. One example of a vegetative measure is the establishment of windbreaks to protect sites vulnerable to excessive wind erosion. Alley-cropping and other agroforestry schemes are also used to optimize the site potentials (see below) and, importantly, mitigate the risks of failure of monocultures. Fallow periods that are sufficiently long enough to allow for the recovery of the soil's fertility could be necessary. Engineering measures implemented for the same general purpose include the construction of bench or broad-based terraces, contour ditches and gully-control structures and protected waterways. Water-harvesting techniques, water spreading and localized irrigation systems are also considered where appropriate to augment the amount of water that is available for agricultural uses.

A key to sustaining and/or improving livestock production on a watershed is retaining an adequate forage cover. The annual production of native forage species often meets this requirement on a watershed in good condition (Pratt and Gwynne, 1977; Jacobs, 1986; Clary and Webster, 1990; Holechek *et al.*, 1998). On a watershed in poor condition, however, it can become necessary for the livestock producer to remove undesirable (noxious) plants to favour the establishment and growth of more highly valued forage species; improve forage production by seeding of species suitable to the conditions encountered; and/or increase water sources of the livestock by drilling wells, constructing a water-harvesting system or building small impoundments to trap and hold runoff water that would otherwise be lost to the livestock.

Improved reproductive methods, applications of fertilizers to achieve optimal tree growth and other cultural treatments have been incorporated into the intensive management of forests and woodlands of many watersheds in the world. Increased knowledge of growth rates and yields of trees has allowed cutting cycles to be specified for the sustainable and/or increased use of many tree species for construction materials, poles and posts, and other wood products. Appropriate prevention measures to protect against wildfire, disease and insect infestation are also known in many instances, with their applications necessary in sustaining the array of benefits obtained from these forests and woodlands. People are recognizing the need to invest their labour, time and other resources in implementing these measures.

Maintaining vegetative ground cover and preventing or minimizing surface soil compaction are the most important means of sustaining or increasing on-site productivity and, at the same time, reducing the loss of soil due to surface erosion. By doing so, raindrop energy at the soil surface is reduced as is runoff of water over the soil surface. In some cases, structural measures are needed that reduce steepness of slope, reduce the length of slope segments along hillslopes, and intercept and direct water to areas where drainage occurs with minimal soil disturbance (Sheng, 1990). A variety of soil conservation measures have been adopted to assist people in meeting the purpose (Box 2.3). These measures focus principally on the reduction in excessive surface runoff and soil loss. In addition to surface soil erosion, soil loss along hillslopes can occur from gully formation and expansion as well as soil mass movements such as hillslope failures, debris flows and mudflows. Watershed measures to prevent gully erosion focus on many of the same management practices in Box 2.3 that reduce surface runoff. Once gullies form, however, structural measures such as gully dams made of loose rock, woody debris, gabions and bioengineering measures are often required. These are costly and require protection for periods of time while vegetative cover is restored on the land. Hillslope failures can occur as a result of natural processes but can be exacerbated by the loss of trees and other woody plants on hillslopes (loss of root strength) and construction of roads along steep slopes. Once soil mass movement occurs, efforts are needed to revegetate and protect sites.

The linkages between the soil conservation measures, the varying forms of soil erosion and downstream consequences of increases in the flow of water, soil erosion and sediment transport are illustrated in Fig. 2.4. What is not illustrated are watershed management practices that reduce stream channel flow and thereby the capacity of streams to transport sediment downstream such as the restoration of riparian corridors and wetlands.

In addition to preventing or minimizing soil erosion from water, there is also concern of wind erosion in many drylands. As with water erosion, a principle management objective to prevent wind erosion is maintaining grass and herbaceous cover of the soil for as long as possible each year. Grazing management, no-till cropping and using ground cover with crops help in this regard. Windbreaks and shelter belt plantings of trees, shrubs and perennial grasses around agricultural fields and pastures also help reduce wind erosion. Wind erosion is particularly problematic during periods of drought in areas where perennial vegetation has been replaced with annual crops.

Box 2.3. Soil conservation measures to reduce surface runoff, soil loss and avoid damage caused by erosion: a few examples. (From FAO, 1977; Sheng, 1990; Brooks *et al.*, 2003; and others.)

Minimizing/reducing surface soil erosion:

- Maintain grass and herbaceous cover of the soil for as long as possible each year through grazing management, no-till cropping, ground cover with crops;
- Locate livestock-watering facilities to minimize runoff to water bodies;
- Avoid logging and heavy grazing on steep slopes;
- Lay out roads and trails so that runoff is not channelized on steep, susceptible areas;
- Maintain or develop vegetated buffers between agricultural fields and water bodies.

Erosion control on steep slopes:

- Furrows and trenches constructed along contours of slopes that intercept surface runoff and promote infiltration and establishment of vegetation on hillslopes;
- Construction of terraces that provide flat soil surfaces conducive to cropping; many are reverse slope benches that promote infiltration and are often used in humid tropics with high volumes and intensities of rainfall.

Gully erosion:

- Establishment of check dams constructed from loose rock, rock bound with wire mesh (gabions), logs and other woody debris, or some combination of these materials that are designed to store and release water, trap sediment and facilitate revegetation in gullies.

Soil mass movement:

- Maintaining trees and other deep-rooted plants on steep hillslopes;
- Avoiding road cuts across, and at the toe of, steep slopes;
- Establishing bioengineering structures (e.g. gabions with tree saplings) to stabilize soil slips and areas susceptible to debris flows;
- Avoiding or minimizing human occupation or development in areas susceptible to landslides, debris flows in channels (e.g. zoning of hazardous areas).

Increasing water yield: implications for water supply

Water supply systems, whether for municipal and industrial use, irrigation or hydro-power, most often involve reservoirs for storing and redistributing surface water for large-scale projects, or pumping of groundwater for local drinking water. The focus in this discussion is on the potential for increasing water yield from watersheds to augment local water supplies, and also to consider implications of land use for larger-scale water supply and distribution systems.

Vegetative changes that reduce evapotranspiration (ET) rates on a watershed generally increase water flow from the watershed because ET accounts for much of the disposition of precipitation falling on watershed landscapes (Satterlund and Adams, 1992; Brooks *et al.*, 2003; Chang, 2003). ET is often reduced by changing the composition, structure or density of the vegetative cover on a watershed (Table 2.2).

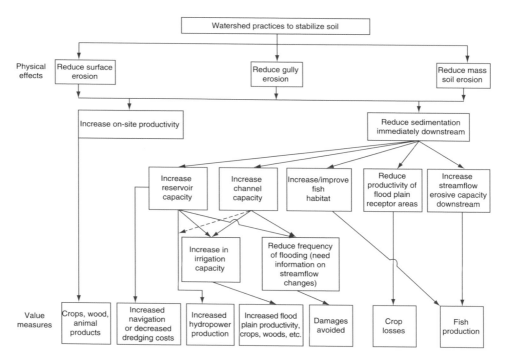

Fig. 2.4. Linkages among soil conservation measures, varying forms of soil erosion and downstream consequences of increases in the flow of water, soil erosion and sediment transport. (Adapted from Gregersen *et al.*, 1987.)

Experimental watershed studies worldwide have shown that water flow can be increased from 5 to 650 mm of pretreatment streamflow regimes (Bosch and Hewlett, 1982; Whitehead and Robinson, 1993; Ice and Stednick, 2004) when:

- Vegetation is converted from deep-rooted plant species to shallow-rooted species;
- A vegetative cover is changed from plant species with high interception capacities to species with lower interception capacities;
- Plant species with high transpiration losses are replaced by species with low transpiration losses.

The magnitude of the increase in water flowing from a watershed following changes in vegetative cover also depends on the amount of annual precipitation, soil characteristics and the percentage of the watershed affected by the change in vegetative cover. Higher water yield responses are generally expected on watersheds with high annual precipitation and deep soils, while responses are lower with the opposite conditions. There is usually a lower limit of annual precipitation below which there would be little or no effect on water yield with changes in forest conditions. Annual precipitation must exceed 450 mm before detectable changes in water yield occur as a result of the above changes in vegetative cover in western USA (Hibbert, 1983). Similar thresholds likely occur elsewhere. The length of time that water yields on the treated watershed continue to exceed pretreatment levels is influenced by the rate at

Table 2.2. Water yield increases following vegetation manipulations: some examples of water yield increases following removal of vegetation.

Vegetative type	Range of water yield increases (mm/year)	References
Conifers and Eucalyptus	200–650	Global studies (Bosch and Hewlett, 1982; Pilgrim, 1982)
Ponderosa pine (*Pinus ponderosa*)	25–165	South-western USA (Hibbert, 1983; Baker, 1986)
Pinyon juniper (*Pinus* spp., *Juniperus* spp.)	0–10	South-west USA (Hibbert, 1983; Baker, 1986)
Hardwoods	60–400	Global studies (Bosch and Hewlett, 1982; Pilgrim, 1982)
Aspen (*Populus tremuloides*)	90–150	USA (Hibbert, 1983; Baker, 1986; Verry, 1986)
Cottonwood (*Populus* spp.)	>1000	Riparian areas in south-west USA (Horton and Campbell, 1974)
Mesquite (*Prosopis* spp.)	300–500	Riparian areas in south-west USA (Horton and Campbell, 1974)
Saltcedar (*Tamarix* spp.)	1000–2200	Lysimeters in south-west USA (Van Hylckama, 1970)
Rangelands		
Sagebrush (*Artemisia* spp.)	0–12	South-west USA (Hibbert, 1983)
Semi-desert shrubs	Negligible	South-west USA (Hibbert, 1983)

which vegetation regrows on the watershed following treatment. Changes that increase vegetative biomass, or conversions from annual crops to perennial vegetation, including forest cover, will have the opposite effects of those portrayed in Table 2.2. In addition, conversions from vegetation cover with lower consumptive use of water to vegetation with higher consumptive use (e.g. converting a hardwood forest to a conifer forest) would be expected to reduce water yield.

The downstream consequences of increasing or decreasing water yields through changes in vegetative cover on a watershed are illustrated in Fig. 2.5 and they depend on several factors. If water yield augmentation at a reservoir was an objective of watershed management and the appropriate conversions of vegetative cover were implemented, the realization of tangible increases in water at the reservoir would depend on:

- Seepage from stream channels, called transmission losses, which link the watershed to the reservoir;
- Evaporation losses in riparian areas, stream channels and the reservoir itself;
- Reservoir storage capacity at the time of water yield increases; for example, if water yield increases occur during the period of time that the reservoir is usually full, no actual increase of water would be captured; possible increases in water flows during the flood season could have adverse effects;
- Soil erosion and sediment loading to the reservoir; if this accompanies a watershed treatment, storage capacity reductions at the reservoir might offset any increases in water yield and thus not warrant the treatment.

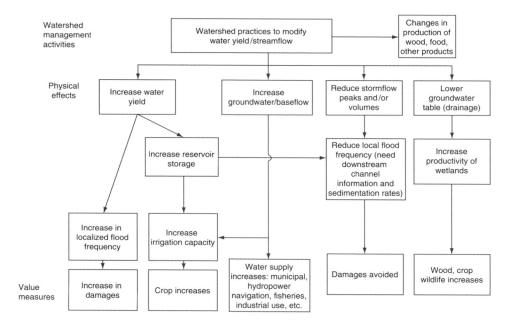

Fig. 2.5. Linkages among watershed practices to modify water yield and streamflow timing and downstream consequences.

In addition to the above, there are often other constraints imposed that restrict widespread conversions of vegetative cover across watersheds such as biodiversity concerns, wildlife habitat loss and impaired water flow regimes that affect fisheries.

Various land-use practices on watersheds can result in changes in the pattern of flow, increasing or decreasing peak flows or low flows. If such changes are of sufficient magnitude or extent, they can have downstream consequences as well, as depicted in Fig. 2.5. Changes in land use on peak flows are discussed later in this chapter; however, low flows (or baseflows) can likewise be altered when there are major changes in vegetative cover across large portions of watersheds. For example, when watersheds undergo major deforestation, low flows usually increase; forestation of croplands or grasslands usually diminishes low flows (Andreassian, 2004). Such changes can affect water quality, aquatic ecosystems and fisheries. Exceptions to this response may occur when such land-use changes occur in cloud forests (see Chapter 5).

Water harvesting is a structural method of augmenting available water supplies in the more arid regions of the world. A typical water-harvesting system consists of a catchment area that is treated to improve runoff efficiency; a smaller storage facility for the harvested water; and a distribution system to transport the stored water to its point of use (Box 2.4). Water spreading by constructing a series of terraces to deflect overland flows of water onto strips of planted crops is a proven method of distributing intermittent water flow onto the landscape to enhance agricultural crop and forage. Water-harvesting and water-spreading methodologies are often used in combination to increase water supplies and on-site productivity on a watershed.

Water that is available following periods of abundance can often be salvaged for later use through a variety of methods that reduce evaporation rates and seepage losses

> **Box 2.4. Water harvesting: some examples of methods and applications.**
>
> Numerous water-harvesting systems have been installed in the arid regions of the world and many more are being installed. Most of these installations have been successful in meeting their stated goals and objectives. A few examples are cited below.
>
> **Western Australia**
>
> Australia was among the first Western countries to install operational water-harvesting systems to provide water for livestock and domestic needs. The Public Works Department of Western Australia initiated this programme of constructing catchments in 1948. The catchments were made by clearing, shaping and contouring to control the length and degree of slope and compacting with the aid of pneumatic rollers (Burdass, 1975). About 2500 catchments averaging 1 ha in size supply water principally for livestock use. Additionally, there are 21 catchments totalling about 700 ha and ranging from about 12 to 70 ha that are used to furnish domestic water for small towns.
>
> **Pakistan**
>
> The Pakistan Forest Institute installed a system of micro-catchments on more than 40 ha in an area of the country that receives 250–300 mm of annual rainfall in 1982. The purpose of this project was to establish plantations of *Acacia*, *Prosopis*, *Tecoma* and *Parkinsonia* tree species on sites receiving less than optimal precipitation amounts (Thames, 1989). Survival of these species was 80–90%, while the survival on (control) sites without micro-catchments was only 10%. *Acacia tortilis* had the highest survival rate and grew more than 1 m annually.
>
> **South-western USA**
>
> A water-harvesting system consisting of a gravity-fed sump, a storage reservoir, a series of catchments and an irrigation system occupies nearly 2 ha of retired farmland near Tucson, Arizona. The combined designed capacity of the sump and reservoir – treated with sodium chloride (NaCl) – is approximately 2400 m^3 of water (Karpiscak *et al.*, 1984). The reservoir was covered with 250,000 empty plastic film cans to decrease evaporation losses. The catchments – also treated with NaCl to decrease infiltration – are used to concentrate rainfall runoff around planted agricultural crops and tree species in untreated planting areas at the base of the catchments.

from impoundments of water on a watershed and downstream (Satterlund and Adams, 1992; Ffolliott *et al.*, 1995a; Brooks *et al.*, 2003; Chang, 2003). Evaporation from small impoundments such as catchments for storing livestock water can be reduced by covering these water bodies with polystyrene, rubber sheeting or floating blocks of wax. Aliphatic alcohols and other liquid chemicals that form monomolecular layers on the water surface have also been used on an experimental basis to reduce evaporation on larger water bodies. Approaches to reducing seepage losses from reservoirs, irrigation laterals and earthen canals constructed on pervious soils include:

- Compaction of the soil within these structures;
- Treatment of the soil surface with chemicals to break up aggregates;
- Lining the canals and bottoms of small reservoirs with impervious materials; however, this method can be too expensive for large reservoirs.

Improving water quality

There can be adequate water to meet local needs, but its quality might be such that it is not suitable for its designated use or uses. The quality of water flowing from upland watersheds is affected by the geological–soil–plant–atmospheric system and land-use practices. Water flowing from a watershed in good condition usually has high physical, chemical and biological quality, and is therefore suitable for a wide array of ultimate uses (Brooks *et al.*, 2003; Chang, 2003). However, the reverse situation is the general case when a watershed is in poor condition.

Effects of land use on water quality must be understood if we are to improve and sustain watershed and stream health. Therefore, it is important to recognize that the hydrologic changes that accompany land use are also important in affecting water quality – a fact that is often overlooked. Hydrologic processes that affect soil erosion, sedimentation and loading of nutrients such as phosphorus are critically important in this regard. Approaches to improve the water quality when this becomes necessary should be targeted to the cause or causes of the problem as illustrated in Fig. 2.6. Implementing the BMPs approach is one way to mitigate non-point pollution and lead to improved water quality (see Chapter 5). This approach involves identification and then imple-

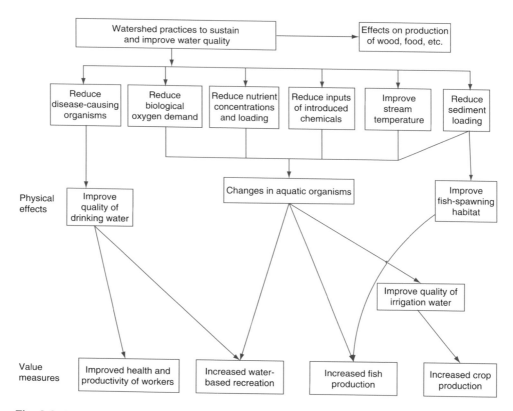

Fig. 2.6. Watershed practices to sustain or improve water quality and their downstream linkages. (Adapted from Gregersen *et al.*, 1987.)

mentation of land-use practices that reduce or prevent non-point pollution and other environmental problems associated with poor water quality (Brown *et al.*, 1993; Brooks *et al.*, 2003). Many BMPs are known for agricultural, forestry and road-construction activities; however, BMPs are not well known for some types of chemical pollutants.

Land-use practices that lead to improved water quality include maintaining riparian (streamside) buffers that can help mitigate sediment and nutrient discharge into streams from adjacent agricultural croplands or livestock-grazing areas. Buffers can either consist of natural riparian vegetation that is left in place next to water bodies or they can be established to achieve specific objectives such as shading small streams to maintain cooler water temperatures. Agroforestry practices offer a potential to provide multiple products for landowners while protecting water bodies with the establishment of buffers, or practices on watersheds that are susceptible to soil erosion, sedimentation problems and nutrient delivery to water bodies such as steep slopes, springs and wet-soil areas.

Two questions that might arise at this point regarding investments of time and other resources in improving water quality are: Why should farmers or livestock producers change their practice to improve water quality? How can they benefit if they made this change? Although there can be economic value in protecting or improving water quality (see Chapter 4), the essence of IWM is that by implementing good land-use practices, benefits accrue to the landowner and/or the user of the land through sustained productivity, i.e. minimal soil erosion and so forth, and a byproduct of such use is that water quality will normally be improved or sustained at high levels. Aquatic ecosystems and fish production are examples of benefits derived from high-quality water. Improved water quality can also benefit outdoor re-creational opportunities that have economic implications to an area.

Mitigating effects of landslides, debris flows and flooding

Conversions of a mountainous watershed from forest to agricultural crops or pastures are often blamed for the occurrences of landslides, debris flows and flooding. Reasons for this thinking are often based on people's perceptions of the hydrologic role of forests rather than scientific evidence. Forests produce relatively low levels of stormflow and, generally, provide greater soil stability than other vegetative types because of their protective vegetative cover, high consumptive use of soil water, high rates of infiltration and high tensile strength of roots. These attributes are beneficial to watersheds in mountainous terrains that are subjected to torrential rainfall (Sidle, 2000). However, there is a limit to the level of protection that a forest cover provides when the vast majority of mountainous watersheds are forested and managed for slope stabilization and torrent control. As the magnitude of rainfall becomes extreme, the extent to which forests can mitigate landslides, debris flows and flooding diminishes.

Land scarcity and human behaviour constrain our ability to cope with landslides, debris flows, floods and related extreme events. Davies (1997) stated that debris flow disasters are the result of 'a natural processes of erosion and sediment motion interacting with human systems'. Where land scarcity concentrates people and their dwellings in hazardous areas on watersheds, disasters will occur whether uplands are fully forested or not. Such is the situation in Taiwan (Box 2.5) with a population density approaching 600/km^2. People living on steep hillslopes, at the mouth of small drainages and on flood plains in this country are vulnerable regardless of the upland watershed conditions. IWM practices, projects and programmes are needed in such instances that address the reality of limited watershed protection and the need for removing people from hazardous areas.

Impacts of landslides, debris flows and flooding events are exacerbated at times by improper land use or management practices on upland watersheds, although the extent to which upland watershed conditions are related to these events is unclear. It is known, however, that the effects of watershed conditions on flooding tend to diminish as the magnitude of the storm producing the events increases (Dunne and Leopold, 1978; Brooks *et al.*, 2003; Andreassian, 2004; Calder, 2005). Hydrologists often conclude that if there is rainfall from a large and relatively short event on soils with low infiltration rates, landslides, debris flows and flooding are possibilities regardless of the land-use and management practices on upland watersheds.

Although there are only a few studies that show the extent to which land-use and watershed changes cause increases in flooding, through basic hydrologic principles (see Chapter 5) we suggest:

- Large floods are the result of major precipitation events.
- Land use that causes excessive runoff and sediment flow can potentially increase the magnitude of medium and smaller floods but have minimal affect on large floods.
- Flood damages and loss of life result largely from people occupying flood plains and other flood-prone areas rather than human-induced flooding through land use and watershed management practices in upland areas.

Groundwater implications

Relationships between land use on a watershed and its impact on groundwater are less certain than the land use–surface water relationships discussed above. As watersheds are the ultimate sources of groundwater recharge, there is a clear linkage between soil types, topographic and geologic conditions of a watershed, and its effectiveness in recharging groundwater. On the other hand, land use that reduces infiltration rates over widespread areas of recharge can create more surface runoff at the expense of recharging groundwater. Land use that reduces ET on the watershed can, conversely, increase the amount of water that is available for recharge. Drainage ditches and tile-lined drainage systems that intercept shallow groundwater can essentially transform groundwater into surface channels, affecting groundwater storage and flow as a conference. Concerns about groundwater reductions are manifold. Of particular interest to people is the potential for reducing groundwater-fed perennial streams and impacts on dry-season streamflow.

Changes in the quality of water that reaches groundwater aquifers are a problem to people in many parts of the world. Wastewater from farmsteads or livestock operations and/or the percolation of water from agricultural croplands can add contaminants or nutrients to groundwater that impact use of groundwater resources. Such problems can be serious where groundwater aquifers are made of limestone, in which case water can move quickly into these aquifers which may be sources of drinking water or important sources of flow to streams or lakes. Many people suffer from unsanitary water supplies that can often be attributed to such problems.

Conjunctive management of surface water and groundwater has recently become a focus of agencies and organization responsible for instituting a holistic approach to water resource management as a consequence of the above issues. Conjunctive water management involves coordinating surface water supplies and their storage with groundwater supplies and the storage capacity of underground aquifers (Blomquist *et al.*, 2004). The goal of conjunctive water management in physical terms is:

- Reducing the impacts to the extremes of flooding and drought and flooding on people;
- Maximizing the availability of water-usable supplies;
- Improving the efficiency of water distribution;
- Protecting the physical, chemical and biological quality of water;
- Sustaining ecological needs and aesthetic and re-creational values.

From an institutional standpoint (see Chapter 3), it is also necessary that thorough considerations and assessments are made of:

- How the responsible institutions are connected;
- How these institutions stimulate changes in water management practices;
- How they facilitate or hinder those changes;
- How they shape the available choices that water managers and organizations make;
- How they affect the outcomes that water users and organizations achieve.

Rehabilitation activities

The functioning of many watersheds around the world has been disrupted largely as the consequence of improper land-use practices or devastating wildfire that have caused watershed conditions to degrade from good to poor conditions. Therefore, practices to rehabilitate a watershed in poor condition to a more productive state can also be the focus of watershed management. Among these rehabilitation activities are:

- Establishing protective vegetative covers on severely eroded or otherwise degraded hillslope sites is common. Grasses and mixtures of grasses, legumes and other herbaceous plants are often seeded from aircraft. Most mixes contain annual plants for a rapidly established cover and perennial plants to establish longer-term protection. Mulching might be spread over the soil surface to protect it from raindrop impact. Other hillslope treatments include tilling, temporary fencing, installation of erosion-control fabric, use of straw wattles and lopping and scattering of slash.
- Developing contour furrows, terraces and similar retention structures along hillslopes to capture surface runoff, hold sediment and establish vegetation is often to the merit of agroforestry practices.
- Gullies and soil mass movement is controlled by constructing upstream check dams with straw bales, downed logs and rocks to restrict the movement of eroded soil into stream channels. Construction of rock-cage dams (gabions) can be an effective but relatively expensive option. Other channel treatments are in-channel debris basins, in-channel debris clearing and streambank armouring.
- Stream-channel restoration and riparian (streamside) revegetation are implemented, and riparian buffers and bioengineering practices are employed to hold streambanks in place.
- Limiting intensive agricultural cropping, livestock grazing, timber harvesting and road construction on sites undergoing, or potentially subject to, excessive soil loss. Road treatments aimed at increasing the water- and sediment-processing capabilities of roads and road surfaces can be important to rehabilitation. Among these road treatments are outsloping, placing gravel on the running surface, placing rocks in ditches, culvert removal or upgrading, armoured stream-crossings and water bars.

Because severely degraded watersheds are difficult and costly to rehabilitate, a management emphasis should be placed on incorporating practices that minimize adverse impacts to soil and water resources in the first place into the initial planning of watershed management practices. By doing so, the chain reactions that occur from an initial loss of soil and water resources through advanced states of degradation in which the overall productivity and usefulness of a watershed landscape is minimal can be adverted.

Watershed health: a dynamic equilibrium

The varying roles of forest and non-forest vegetation on streamflow regimes, soil erosion, hillslope and stream channel stability, and water quality are all interrelated and can be considered in the context of a watershed health, which can be considered in the context of a dynamic equilibrium (Reid, 1993; Sidle, 2000; Brooks *et al.*, 2003). The dynamic nature of a watershed system is a delicate balance between aggrading and degrading forces. Changes in flow patterns and inputs of sediment can alter the dynamics of watershed systems. Hillslope processes, stream stability and the physical and biological nature of aquatic ecosystems can also be affected. An overview of water flow–sedimentation relationships and the linkages between riparian forests and aquatic ecosystems (see below) illustrate the role of vegetation management in watershed system dynamics.

A vegetative cover provides the greatest protection against processes of surface and gully erosion, with the lowest rates of surface erosion commonly associated with forested areas. Deep-rooted tree species also provide the greatest slope stability in areas susceptible to soil mass movement. It follows, therefore, that stream systems within forested watersheds characteristically carry low sediment loads (Brooks *et al.*, 2003; Chang, 2003). However, watershed management activities that change vegetative cover, alter water flow regimes and affect soil protection can result in accelerated soil erosion and delivery of sediment into stream channels. The consequences of these activities can be a change in the dynamic equilibrium of the water flow–sedimentation relationships that ultimately affect the stream channel itself.

Changes in streamflow patterns – particularly through engineering diversions – can adversely affect riparian forests and aquatic ecosystems. The need for a better understanding of how changes in streamflow and sediment affect stream channel stability and the long-term consequences on stream systems has led to a stream classification system that is applied in western USA (Rosgen, 1994, 1996; Brooks *et al.*, 2003). This hierarchical stream classification system is based on varying levels of inventory resolution ranging from very broad morphological characteristics to reach-specific descriptions. However, because morphological characteristics considered often change within short distances along a stream channel, the stream types classified by the system apply only to those reaches of the channel selected for classification. The importance of maintaining minimum flows within stream channels and the need for understanding how human activities affect streams on both forested and non-forested watersheds are emphasized in this classification system.

Recognizing the function of riparian vegetation in terms of maintaining stream stability and aquatic habitats is receiving greater attention than in the past (Verry *et al.*, 2000; Brooks *et al.*, 2003; Baker *et al.*, 2004). The role of large woody debris on stream channel morphology, and the effects of grazing livestock, harvesting timber and other human-induced activities on these relationships are increasingly emphasized by watershed managers. Natural processes that cause large trees to become wedged within stream channels play an important role in sustaining aquatic habitats and stabilizing streambanks against the *flashy nature* of many upland streams. The role of upland forest and riparian vegetation and watershed conditions on aquatic habitats and fisheries is an important aspect of forest hydrology and watershed management.

Wetlands and other areas with excess water are often drained for purposes of making the land suitable for cultivation of crops. Groundwater table elevations are lowered. Most wetlands are *expressions* of groundwater that reaches the soil surface, and therefore represent areas on a watershed that stores and then releases water principally by potential ET throughout the year. Extensive drainage of wetlands with tile drains and ditches can convert groundwater more quickly into surface channels, with the potential of altering the dynamic equilibrium of streamflow and sediment relationships in channels that can then cause channel degradation.

Upstream–downstream connections

Upstream–downstream connections have been discussed throughout this chapter and elsewhere in this book. These connections can represent beneficial or detrimental impacts on people depending largely on the nature of the land-use practices on upland watersheds and throughout large river basins (see Chapter 1). Estimates of the nature of these connections and their effects are necessary for economic and environmental evaluations of watershed management practices to determine the impacts of these activities into the future. A discussion of some of these connections follows.

Increases in water flowing downstream from upland watersheds are likely to diminish in magnitude as the distance increases between the upland watershed and downstream sites where the water is used (Box 2.6). These reductions in water flow result largely from transmission losses in channel systems, evaporation of water en route and transpiration by vegetation along the streambank.

Downstream sedimentation results not only from upland soil erosion but also from channel erosion and the mass movement of soil directly into the stream channel or a downstream reservoir. Morphological characteristics of the channel and distances between the affected upland watersheds and downstream reservoirs also determine the quantities and timing of sediment delivery downstream. While the concentrations of sediment might be low at the outset, these concentrations can increase significantly as water flows through the tributaries of a larger river basin and, ultimately, to the downstream places of use. Similarly, the flows of high-quality water that are associated with a well-managed watershed can deteriorate in quality within a larger river basin when other watersheds within the basin are improperly managed.

Cumulative Effects

Land-use and management activities across a watershed can bring about changes in the vegetative, soil and water resources that affect the long-term productivities of watersheds. Multiple land-use activities across the watershed can bring about a variety of changes that might not seem important when viewed in isolation. However, in aggregate, they can affect the quantity and quality of water flowing from watersheds and impacting people living downstream. As mentioned above, many watersheds in the world are subjected to agricultural cultivation, livestock grazing, timber harvesting and other forms of human interventions. People might also alter their land use to obtain water resource benefits. The cumulative hydrologic effects of these land-use practices can have many of the intended but also unintended effects on people living upstream as

---

Box 2.6. Transmission losses of water flowing to downstream point of use: a case study.

A question that should be asked by water resources planners and watershed managers is: How much of an observed increase in water flows obtained by changing the vegetation on an upstream watershed eventually reaches downstream reservoirs and other points of use? In a study in south-western USA, Brown and Fogel (1987) estimated that the increase in water flowing from upstream watersheds in the Salt-Verde River Basin resulting from changes in vegetation on the upstream watersheds was reduced by nearly one-half by the time water travelled 150 km to downstream reservoirs. Transmission losses in the stream channel systems, evaporation of water en route and transpiration by vegetation along the streambank were the causes for the reduction.

---

well as downstream. When the improper land-use and management activities outweigh good land use and management on a watershed, the changes can lead to:

- Excessive surface runoff;
- Increased soil compaction and surface erosion;
- Increased gully erosion and mass soil movement;
- Increased sedimentation in downstream channels;
- Increased temperatures of stream waters;
- Increased export of nutrients.

Possibilities of downstream flooding also increase if surface runoff and downstream sedimentation increase as a consequence of improper watershed management practices. Soil erosion and the export of nutrients reduce the nutrient capital on site and, as a consequence, productivities of the natural resource on these watersheds (Dunne and Leopold, 1978; Satterlund and Adams, 1992; Brooks *et al.*, 2003; Chang, 2003). Increased streamwater temperatures and increases in nitrate, phosphorus and other nutrients detrimentally affect aquatic organisms. The introduction of residues from timber harvesting (tops of trees, branches, etc.) into stream systems can lead to higher levels of biochemical oxygen demands and reduce dissolved oxygen, also adversely affecting aquatic ecosystems.

Therefore, while detrimental impacts resulting from land-use practices on a watershed are always possible, they can be minimized with properly planned and carefully implemented watershed management practices. For example, successful revegetation of erodible sites helps to return watershed landscapes to their former conditions by controlling soil erosion.

When we consider an array of land-use activities across a watershed landscape, recognizing the cumulative hydrologic response is often not straightforward. Cumulative watershed effects can often be initially transparent but could have significant hydrologic effects over time. Furthermore, our ability to take information from local sites and scale up to large watersheds or river basins adds considerable uncertainty to planning and management. Scale and temporal questions and our ability to know thresholds and limits challenge the sustainable use and management of land and water.

If we do a visual assessment for the first time on two different watersheds, on one we see a large-scale forest products processing plant discarding some chemicals and other waste material into the main river coming through the watershed. This observation might suggest that this is a serious problem – much more so than on the other watershed where we only see a couple of small, innocent-looking wood-processing facilities discharging small amounts of pollutants into the river system. The concept of cumulative effects comes into play here. Let us say that on the first watershed there are three large plants producing a combined output of 8000 m^3 of product, valued at $3.4 million, and with a pollution cost of $1.1 million. On the other watershed we find after further study that there are about 800 plants on the watershed which are in the aggregate producing approximately 6500 m^3 of product valued at $2.8 million with a pollution cost (reflecting severity of pollution) of also $1.1 million. We can see that the three large production units produce more income for a lower pollution cost per unit. Thus, since the outputs and incomes matter in the broader IWM context, the situation would be judged more favourably towards the watershed with the three large plants. Small is not always necessarily better, particularly if the cost and effectiveness of water pollution control is much greater for the smaller units because of their dispersion and the lack of economies of scale in control efforts compared with those found in the case of the three larger entities.

3 Institutional Context

Integrated Water Resources Management: The coordinated development and management of water, land and related natural resources to maximize the resultant economic and social welfare in an equitable manner without compromising the sustainability of vital ecosystems.

This is the stated objective of integrated water resources management, adopted by the World Summit on Sustainable Development in Johannesburg in 2002 as part of the wider international strategy for the Millennium Development Goals (MDGs). The concept marks the latest in the evolution of water governance frameworks developed since the 1992 International Conference on Water. The conference established three key principles for good governance (UNDP, 2006):

1. The *ecological principle* for integrating water management around river basins rather than independent institutional users, with land and water governance integrated for environmental reasons.
2. The *institutional principle* for basing resource management on dialogue among all stakeholders through transparent and accountable institutions governed by the principle of subsidiarity – the devolution of authority to the lowest appropriate level, from user groups at the base to local government and river basin bodies.
3. The *economic principle* for making more use of incentives and market-based principles to improve the efficiency of water as an increasingly scarce resource.

Many factors are converging to cause natural resource managers, researchers, decision makers and society as a whole to look increasingly to integrated water resources management, or what we call IWM, as a participatory, practical approach for addressing a wide range of land and water-related problems in the context of the basic principles related to ecology, institutions and economics, as mentioned above. Managing water resources at the watershed scale – although difficult – offers the potential for balancing the many, sometimes competing, demands that people place on water and other watershed resources. The IWM approach acknowledges linkages between uplands and downstream areas, as well as between surface and groundwater resources and reduces the chances that attempts to solve problems in one realm, which will cause problems in others. This approach is an integrative way of thinking about the various human activities that occur on a given area of land (the watershed) and have effects on, or are affected by, water. With this perspective, people can plan long-term, sustainable solutions to many problems regarding land, water and other natural resources. Furthermore, people can often find a better balance between meeting today's needs and leaving a sound resource legacy for generations to come (CWM/NRC, 1999).

The USA is not alone in promoting the IWM approach. Countries belonging to the Organization for Economic Cooperation and Development (OECD) – 30 of the most highly developed countries in the world including the USA – have proposed an environmental strategy, which recommends that the member countries should apply the *ecosystem approach* to the management of freshwater resources and the contributing watersheds (OECD, 2006). This strategy also calls for legal frameworks supported by appropriate policy instruments to ensure the sustainable use of freshwater resources and measures to enhance their efficient use. In other words, this recommended approach to land and water management at the landscape level is an integrated one of many technical and social sciences, many interacting and (hopefully) coordinated laws and many interactions among stakeholder groups with varying interests in the management of watersheds, water and beyond. Therefore, almost by definition, watershed and water management exists within a complex institutional environment.

The issues associated with integrated watershed and water management are not trivial. Without effective planning and management – and without establishing institutions that can be effective in reaching the poor – it is estimated that as many as 139 million people will die from water-related diseases by 2020. Even if the MDGs for water are reached, which is unlikely with current international commitments, between 34 and 76 million people will perish from water-related diseases by 2020. This problem is one of the most critical public health crises facing the world and deserves far more attention and resources than it has received so far (Gleick, 2002).

The economic advancement of India (Sengupta, 2006) and many other countries depends directly on how these countries handle their mounting water crises. The problems are no less serious at the international level as graphically illustrated in the recent book entitled *Water Wars* (Raines Ward, 2002). Avoidance of many potentially serious problems depends largely on the people of the world reforming water governance and moving ahead with programmes of water and land management that prevent the degradation of natural resources and build up the capacity to use resources wisely. This chapter discusses the institutional context for such programmes. It begins with a discussion of water governance – the broad umbrella that embraces the institutional context – and then moves on to discuss specific elements within that context in the following sections.

Water Governance

Governance is about effectively implementing socially acceptable allocation and regulation and is therefore intensely political. Governance is a more inclusive concept than government per se. It embraces the relationship between a society and its government. Governance generally involves mediating behaviour through values, norms and, where needed, through laws and regulations. National sovereignty, social values and political ideology can have a strong impact on attempts to formally change governance arrangements related to the water sector, as is the case, for example, with land and water rights or corruption in natural resource use (Rogers and Hall, 2003). The World Bank has estimated that the lost revenue to governments of the world from illegal logging amounts to between $10 and $15 billion. These losses are substantial in and of themselves, but do not include the environmental losses associated with loss of tree cover. In terms of impacts on watershed functions, such losses can be substantial.

Governance is therefore a term used to describe the whole institutional setting within which daily activity takes place. In one sense, watershed and water governance is inwardly focused on all the nuances of politics, legislative backgrounds, social and economic conditions, as well as organizational landscapes that define a local community, region, country or group of countries. In another sense, governance has to be outwardly focused on the interrelationships between the different political units that share common watersheds, which affect each other through what happens on the watersheds in terms of land and water use. Unfortunately, too often, smaller political units such as counties let politics overshadow the broader good that can come from cooperation among political units.

The more the governments and society accept and promote IWM, the more likely will there be need for changes in the relationships between the public and private institutions that determine the governance of land, water and other natural resources. Therefore, as IWM is endorsed, many countries and/or regions pass through various forms of water reform. One example illustrating water reform and the lessons learned from the reform comes from Queensland, Australia (Box 3.1). Each state, country or region will have different circumstances and, therefore, different reforms that are

---

Box 3.1. Australia: implementing water reform in Queensland. (From Global Water Partnership (GWP). Toolbox case study No. 24.) (www.gwptoolbox.org/)n.d.

A series of legislative and policy developments to reform the water sector in the State of Queensland, Australia, were put in place over 1999–2001 (and ongoing), following Commonwealth (national) government water reform initiatives in 1996. The measures include:

- Use of consultation across the stakeholder spectrum from high level of government through to farmers to help develop plans;
- Preparation of draft policy papers then Bills used to drive process;
- Preparation of supporting legislation for regulation of service providers, reform of water authorities; introduction of third party enforcement for offences, compliance notices, increased penalties;
- Introduction of legislation to enshrine environmental flow requirements in the Development of Water (Allocation and Management) Bill;
- Use of a 'whole of river basin' strategic plan approach within which local resource operation plans are prepared and implemented;
- Integration of the reforms with the local planning processes of Queensland.

Lessons learned

In the process it was felt that:

- An incremental approach, with water planning developing in 'bite-sized chunks', allowed the government to be flexible in response to changing circumstances.
- However, the process would have been streamlined had action been taken earlier to separate regulatory functions from supply or service provision roles.
- Furthermore, a clearer definition of roles and responsibilities should have been done earlier.
- In water allocation to local governments (and, presumably, to other users), the government should not mandate how the allocated water is to be used. Instead, it should limit itself to the allocation, and allow the local governments to specify how the allocated water is to be used.

---

needed. Such reform, however, is a first step in moving toward a more integrated approach to watershed and water management.

Institutional Effectiveness

Chapter 2 has indicated that many of the most useful watershed and water management approaches also involve complex systems of activities and events from a technical perspective. The same holds for the institutional context within which such activities take place. We use the term *institutions* in the broadest sense of its meaning – namely to include all the ways by which people come together to cooperate, coordinate and guide their activities. Therefore, this term includes organizations that people establish, laws that people pass and implement through regulations and policies, and the various forms of collective behaviour associated with social, economic and political mechanisms that people adopt. In other words, at any point in time, there is an institutional context within which watershed management takes place.

A number of institutional mechanisms exist to guide activities on a watershed including those that affect the condition of the watershed, the timing of flows of water, the uses of this water and the settling of disputes over the uses of water. There exists, in most cases, a complex set of intertwined laws and customary rights that affect a watershed and the water flow through and from it. There also exists, in most cases, a variety of organizations from the public and private sectors and groups of stakeholders that determine the uses of a watershed or river basin and the water that passes through it from upstream sites to the ocean where the watershed or river basin ends.

If all is in harmony on a watershed, life goes on with no issues arising. People use land, water and other natural resources as they see fit and everyone is largely satisfied with the way the watershed or river basin is managed. Unfortunately, this idyllic situation does not exist. Many of the complications introduced into the *real world* in an institutional context relate to the following:

- There are complications introduced by the fact that watershed boundaries are defined by physical factors and seldom coincide with political boundaries. This becomes a complicating factor if a river flows through several countries and international treaties need to be negotiated with respect to the management and use of the water in the river. A similar situation is often present in the case of interstate, interprovincial or intercanton river basins.
- Stakeholder groups normally include a highly diverse set of entities from government agencies, quasi-government entities, NGOs, the private sector and various groups of people representing the interests of indigenous groups, environmentalists and others. The interactions among these groups can be complex if they truly form partnerships that move toward a common set of goals and objectives for the watershed or river basin in question and plans for reaching the milestones set by these groups.
- Watershed and water management responsibilities are typically shared by a multitude of organizations including those with jurisdiction over land management, water management, forestry activities, agricultural extension, rural development, transportation and so forth. While no single agency usually has complete coordination responsibility or authority in these situations, there are exceptions.

- Complexities are also introduced by the general lack of market prices to guide choices among activities and priorities that often conflict. Market institutions do not exist for most activities involved in managing and protecting a watershed or river basin, although many societies are currently moving toward use of markets and quasi-market mechanisms.
- There generally is an intricate set of intertwined and sometimes conflicting laws and policies that have emerged through time to govern behaviour on a watershed or river basin – i.e. what can and cannot be done, what has to be done, who has rights to the services and goods from the watershed and who has responsibility for these various actions.
- Policies are sometimes created that hinder watershed management authorities. This hindrance is generally not done intentionally. Rather, it is likely that watershed and water management goals, objectives and benefits might not be considered, not understood or are given low priority. For example, resettlement policies for displaced people tend to be driven by forces operating at a great distance from upland watersheds – forces that carry more political clout than those concerned with watershed management and conservation. Resettlement often occurs with inadequate regard to whether the land can sustainably support those being settled and whether the newly arrived settlers will create watershed problems downstream.
- Policies that affect watershed management and use in the uplands can be vaguely defined, inadequate in scope or essentially non-existent – particularly in articulating the desired relationships among the people living both upstream and downstream. Nevertheless, upstream activity can have a significant cumulative effect downstream where the larger populations often live.

In what follows, we take these institutional issues into account and identify some of the ways in which institutions have been created to address these issues. First, we will describe the *typical* institutional context for watershed management and some of the variations that exist. Second, we will look at the conflicting interests of varying groups of stakeholders and some of the approaches that have been tried to deal with the trade-offs among conflicting interests and the lessons learned. If conflicts did not exist, that is, if all stakeholders on a watershed had common interests, there would be little interest in implementing IWM. It is the existence of conflicting interests and the need for negotiated trade-offs among interests that makes a conscious approach to IWM and the related policy making so important – indeed crucial in some situations.

Dealing with Conflicting Interests

A large number of laws, regulations and policies exist in the USA and other *developed countries* that guide the effectiveness with which the responses to increasing demands and needs for water and other natural resources are implemented. With specific respect to policies, proposals generally move to legislatures for laws and to responsible administrators for regulations if the issues of concern are a matter of implementation of policies already in place. Such policies are implemented through four main mechanisms – the promotion of local commitment and participation, legal and regulatory mechanisms, fiscal and financial mechanisms to influence private behaviour, as well as public investment and improved management of resources. A combination of these mechanisms is often the most effective for policy implementation.

However, emerging issues cannot be resolved at the administrative level within the existing policy framework in some cases. These issues are likely to emerge as policy concerns requiring administrators to review the nature of the existing policies to determine whether the problem confronted is because these policies are ineffective, and, therefore, the formulation of more effective policies is necessary. Issue assessment, policy design and related operational actions take in a social and institutional setting that is unique for every country. This uniqueness relates to differences in organizations, customs, laws, regulations, rights, responsibilities and informal rules that guide and influence the success or failure of a particular policy or action. Effective policy actions can require changing institutions and/or developing new policy instruments.

An overview of issue assessment, policy review and policy design when necessary, and assessing organizational feasibility for implementation of existing and/or new policies is presented in Annex 3.1 and further discussed in chapter 4 dealing with planning and policy formation.

Institutional arrangements specify who benefits from water use and establish incentives that guide water use. Well-designed and functional institutional arrangements can set up regulations, pricing mechanisms, water rights and government interventions to guide the effectiveness of water use. However, inadequate institutional arrangements can impede efficient water use and cause serious problems of waste and misuse.

On most watershed landscapes, people can have conflicting ideas on how the land, water and other natural resources should be used ranging along a continuum from the extreme view of total protection of these resources for future generations to the other end of the spectrum where the view is to use the resources and not necessarily worry about future generations. Therefore, the sustainability issue is key in establishing trade-offs in the use of watershed-based resources. There are also spatial issues that enter the picture. What happens on upstream watersheds affects people living downstream in large river basins cutting across several countries. That is, what one community or landowner does upstream affects the quantity and quality of water available for communities and landowners downstream. On almost any upstream or downstream area, there are conflicts among agriculturalists, industrial entities, environmental groups and landowners (household needs for land, water and other natural resources). Importantly, groundwater issues also figure prominently in discussions of watershed management. Some countries actually recharge underground aquifers from surface runoff or river water that would otherwise pass through the watershed or river basin unused except for in-stream use by living organisms.

Many institutional mechanisms have been developed to deal with the various conflicting interests in resource use in a watershed context. These mechanisms include customary rights, laws, treaties and incentives and market-based institutional mechanisms such as creating effective markets for water and other natural resources. The effectiveness of these institutional mechanisms (in turn) depends on the nature of the land and water management and user organizations that already exist or are created. We look at each of these in the following sections.

Customary Water Rights, Water Laws and Treaties

With the complexities associated with the integrated management of natural resource use and conservation within a watershed or river basin, it is not surprising that there exists, in most countries, an intricate web of customary water rights, laws, regulations and policies that need to be considered and harmonized. Some of these concerns relate to land tenure rights and water rights; some are associated with local government affairs while others to regional or state actions; and still others to countries and their relationships in terms of the management and use of joint water bodies or groundwater aquifers.

Customary and statutory water rights

When there was plenty of water for everyone, customary water rights were often enough to provide an orderly picture of who had the rights to what water and when. However, as water has become more scarce, with increasing populations of people and increased per capita use through expanded economic development, in-stream flow requirements for the environment, agricultural production through irrigation and other forms of use, so has the need for more formal laws that can be brought to a court of law when necessary. Customary water rights can be a major barrier to successful establishment of effective statutory laws and regulations regarding water rights in many parts of the world, because such laws and rules are based on long-standing practices that are not codified in written form but very much a part of the code of conduct of local people. These habits and rights become difficult to change. Effective statutory legislation depends not only on its creation by legislatures but also on its implementation and enforcement. These last two factors can be difficult to resolve when customary water rights differ significantly from the legislated rights under the new law. As Burchi (2005) says 'failure to recognize the existence and resilience of customary practices, and to take them into account in "modern" water resources legislation, is a recipe for social tension'.

Customary water rights result from the customs and practices of indigenous groups of people. They are rights that are recognized by these groups as being binding upon them. In some countries such as Canada, customary rights are recognized in the constitution of the country. In other countries, they have a much more shadowy existence. Customary water rights are often rooted in customary land law that governs the use of the land. Water rights went along with land ownership or control. Customary water rights in Canada have been implied by the courts in aboriginal title to land, in treaty-based land rights, in land reserve rights and in common law land-based riparian rights (Burchi [2005] based on a study by Nowlan [2004]). Burchi goes on to point out that these features of customary water rights are not devoid of legal implications when it comes to their interface with statutory water rights and with modern water legislation which severs the land–water link and decouples water rights from the land. This point is discussed later in this chapter.

A key point about customary water rights is that *modern* water legislation needs to consider customary rights if new legislation is to stand a chance of effective implementation. There will obviously be a transition period during which an adjustment will take place from customary to statutory law for water use and disposal and customary to statutory law that is related to management of the land over which water flows and under which water exists in groundwater aquifers. Four case studies were commissioned by the Food and Agriculture Organization (FAO) of the United Nations (UN) and the International Union for Conservation of Nature (IUCN) in a joint research project investigating the interface of customary and statutory water rights to gain insight on this issue. More specifically, customary water law and practices in Guyana (Janki, 2004), Nigeria (Kuruk, 2004), Canada (Nowlan, 2004) and Ghana (Sarpong, 2004) were investigated, analysed and compared in terms of their respective effectiveness. These case studies illustrate the complexities that exist in moving from customary to statutory law concerning water and the difficulties that arise when statutory law attempts to integrate with customary law or rights.

Land tenure and water rights

Basic to any discussion of institutional issues and mechanisms in IWM is an understanding of the nature of the land tenure rights on the one hand and water rights on the other. While the two forms of tenure are both types of legal rights that can be asserted in a court of law against third parties including the state, and share the same purposes in that they permit society to make an orderly and legal allocation of valuable resources and confer security to the owners of the rights to encourage investment, they also differ in many respects. In contrast to land tenure rights where the resource itself might be privately owned, water originating from natural sources typically remains under state ownership or control. Water rights are also increasingly time bound, while land tenure rights tend to be unlimited. Although private land tenure rights tend to be freely tradable, water rights are seldom tradable and are highly regulated when traded (Hodgson, 2003). Furthermore, while international law plays a key role in terms of water rights in many situations, it seldom has a role to play in terms of private land tenure rights. Hodgson points out that because of these differences and because there are few formal links between land tenure regimes and water rights regimes, water law has become a discipline of its own.

The question arises, therefore, as to whether this issue matters in the context of the increased emphasis on integrated approaches to land and water management in an integrated watershed framework. Hodgson (2003) concludes that it probably does not matter so long as both rights' regimes are clear and there is a detailed understanding of the relationship between the two, for example, in terms of how water rights affect the value of land tenure rights and the sale of such rights. In cases where customary law has prevailed, problems can arise unless the rights are more clearly articulated and established on a legal basis.

In the case of surface water, countries have varying forms of legislation that regulate land tenure rights based on the common knowledge that what is done on a piece of land upstream can affect downstream users in terms of the quantity of water that passes downstream, the timing of these water flows and the quality of water available. However, a special area of concern is the linkage between groundwater rights and land tenure rights. Groundwater is a key resource in most parts of the world. About 75% of the drinking water in Europe comes from the groundwater, whereas it is as high as 98% in Denmark. In the USA, nearly 50% of all drinking water supplies come from groundwater resources, with about 97% in rural areas (Burchi, 1999). There is no simple answer to how best groundwater rights can be linked to land tenure rights. Countries and in-country states, provinces or cantons are experimenting with different regulatory approaches including ones coupled with the establishment of groundwater districts – which are a form of water-user association – and other community-based approaches to manage the use of groundwater resources (Hodgson, 2004).

Regardless of the approach chosen, a comparative analysis of the groundwater legislation passed in the 1990s showed that groundwater is rapidly losing the *intense* private property connotation it traditionally had and that user rights in it are no longer accrued from ownership of overlying land but from a grant of the government or of the courts (Burchi, 1999). The public domain status of groundwater underpins the usufructuary nature of individual groundwater rights and the authority of the government to grant such rights.

In-country water sharing agreements

Issues become complicated when a river flows through a number of states, provinces or cantons in-country. They become even more complex when the river flows out of a country into another (Wolf, 2006). Such is the case with the Colorado River in the western USA. The Colorado River Compact of 1922 separated the river basin into the Upper and Lower basins and apportioned the water flowing in the Colorado River to the respective states in the two basins. This compact also stipulated that the Upper basin states could not withhold water from the states in the Lower basin; that the Lower basin states shall not require the delivery of water which cannot be *reasonably* applied to domestic and/or agricultural uses; and that water to flow into Mexico must come from any surplus waters not allocated to the states by the compact or if this surplus is insufficient, the *burden* of the deficiency will be shared equally by the Upper and Lower basins.

The Boulder Canyon Act of 1928 authorized the construction of Hoover Dam on the Colorado River and the All-American Canal in the Imperial Valley in California. It also apportioned the Lower basin states' allocation of water specified in the Colorado River Compact. A dispute between California and Arizona over rights of the two states to the use of water specified in the Boulder Canyon Act was settled by the US Supreme Court in 1963 – California had been diverting more than its share of Colorado River water. Much to California's dismay, the Supreme Court ruled that the allocations of water outlined in the Boulder Canyon Act would be enforced. This ruling also led to the approval of the Central Arizona Project in 1968, a project that diverts water from the Colorado River for delivery into central Arizona to alleviate water shortages.

The treaty with Mexico into which the Colorado River flows before entering the Gulf of California guaranteed the delivery of 1.5 million acre-feet of water into Mexico except in events of *extraordinary drought* or *serious accident* and up to 1.7 million acre-feet in the years of surplus water in the Colorado River. Within the operating criteria for the Colorado River specified in the 1968 agreement that authorized the Central Arizona Project was a stipulation that water would be released from the reservoir behind Hoover Dam to meet the Mexican treaty obligations when necessary. With respect to water quality, because the Colorado River water entering Mexico had usually exceeded the agreed salinity standards, the USA was (in effect) violating the treaty with Mexico. To remedy this situation, the Colorado River Basin Salinity Control Act was signed by the two countries in 1974 to adopt measures to ensure that the delivery of water into Mexico had an average salinity of 115 parts per million or less – an agreed to level. The USA also financed salinity control projects to remove excess salts from the Colorado River water before it entered Mexico.

A large number of states in the USA have also entered into in-country agreements among themselves and with agencies in the federal government relative to the use, management and sustainability of the waters flowing in the Missouri, Mississippi, Ohio and other large river basins. To elaborate on these agreements in detail is beyond the scope of this chapter. However, they address issues such as maintaining navigation for commercial shipping, the development of hydropower facilities, diversions of water flow in the rivers for industrial, agricultural and domestic uses, as well as fishing rights of the bordering states.

International treaties and transcountry boundary institutions

As stated on numerous occasions in this book, what people do upstream affects people downstream. In some cases, governance mechanisms in the form of laws are established to limit what upstream land and water users – whether public agencies or private individuals, communities or corporations – can do in the context of how that affects downstream landowners and users. Where several countries are located within a river basin, the issues and governance mechanisms become much more complex and are often resolved through international treaties, agreements or compacts. A few examples of these treaties and the *administrative institutions* often established as part of the treaties are presented below to illustrate the diversity of issues considered and procedures for resolving conflicts among the parties to the agreements when specified (Wolf, 2006).

A Committee for Co-ordination of Investigations on the Lower Mekong River was established in 1957 by Cambodia, Laos, Thailand and Vietnam in response to a directive of the UN Economic Commission for Asia and the Near East. This committee was created to provide technical assistance and supervise the planning of water development projects in the lower Mekong River basin. Procedures for conflict resolution were not specified and, unfortunately, little was accomplished because of subsequent conflicts in the region. A convention was later signed by Laos and Thailand in 1965 to interconnect the electric grids of two hydropower plants dependent on the flow of water in the Mekong River. Once again, regional conflicts hindered effective execution of this convention at the time of its signing. An agreement to create a Joint Committee to prepare, propose and establish *rules* for the utilization of river water among countries within the Mekong River basin and interbasin diversions of water when necessary in the wet and dry seasons was authorized by Cambodia, Laos, Thailand and Vietnam in 1995. Three successive stages of conflict resolution were stated in this treaty. First, a conflict is taken before the Joint Committee for hearing. Second, if the conflict persists, it is referred to the governments represented by the Joint Committee. Finally, these governments can request mediation of the conflict.

An early treaty between Great Britain and Ethiopia in 1902 relative to water flows in the Nile River stipulated that Ethiopia would not allow construction to be undertaken on the river that might reduce the *natural flow* of the Nile River unless Great Britain and Sudan agreed. Ethiopia also ceded a tract of riverfront land for use in a railway connecting Sudan with Uganda. Conflict resolutions were not specifically addressed in this agreement. A series of subsequent droughts from 1913 to 1922 spurred a later agreement between Great Britain and Egypt relative to the use of water in the Nile River for irrigation purposes. Navigation concerns were also considered in this 1929 treaty and Egypt and Sudan were required to agree before any construction could take place to increase local water supplies. Conflict resolutions and arrangements stipulated that a third party be included in the negotiations. Technical Committees established in a 1950 treaty between the United Arab Republic (Egypt and Syria) and Sudan authorized that flow reductions in the Nile River – when they occurred – would be shared equally between the countries. Efforts to reduce the large evaporation losses in Sudanese swamps were also initiated. Conflicts were to be brought before a council for resolution. A bilateral framework to establish cooperation between Egypt and Ethiopia on the use of Nile River waters was formulated in 1993. These two countries were required in this compact to negotiate conflicts when they arose, although specifications on the details of resolving the conflicts were not included in the agreement.

As one might expect, there has been a large number of treaties in relation to the shared use of water in the historic Danube River. These agreements addressed a variety of concerns raised by the bordering countries including navigation on the river, flood control and relief measures, hydropower and hydroelectricity generation and protection of water quality. A snapshot of some of these treaties is presented below to illustrate this diversity of issues addressed.

A multilateral treaty authorized by most of the countries bordering the Danube River in 1921 established a commission to review all of the proposed *hydraulic works* in the river and allow implementation of only those not interfering with free navigation. A technical council was charged with resolving conflicts that might arise. Various bilateral agreements were later signed by the countries bordering the Danube River in the 1940s and early 1950s. These mostly bilateral treaties focused largely on flood control and relief measures and the respective country's fishing industries. Various commissions and councils were charged with issues of conflict resolution. Italy and Switzerland agreed to a convention concerning the protection of Italo–Swiss waters against pollution in 1972, with a commission created to enforce the protection of both surface and groundwater resources. Procedures for conflict resolution were not specified.

Recently, a treaty between the Czech and Slovak Republics centered on technical cooperation and assistance in the areas of water economy, agriculture and food industries and forestry activities within the Danube River Basin. One Article in this 1992 agreement mentioned *joint measures* for the administration and utilization of common bodies of water and the control of floods. An agreement between Moldova and the Ukraine relative to the prevention, control and reduction of solid, liquid and gaseous substances in transboundary waters was signed in 1994. Procedures for possible conflict resolution were not specified. A convention that established an International Joint Commission for the protection of the Danube River was signed by most of the bordering countries in 1994. The focus of this convention was placed on sustainable and equitable water management including the conservation, improvement and use of water, regulating flooding events, hydropower production, as well as water transfers and withdrawals. Negotiation with the assistance of the International Joint Commission was specified as the key means of conflict resolution with submission to the International Court of Justice for arbitration when necessary.

Incentives and market-based institutions

The use of economic instruments in dealing with institutional issues relating to watershed and water management is increasing, but has far from reached its full potential. Until recently, most governments have relied primarily on direct regulation in watershed and water resources management. However, economic tools can offer several advantages such as providing incentives to change behaviour, raising revenue to help finance necessary investments and/or establishing user priorities and achieving management objectives at the least possible overall cost to society. Prerequisites for a successful application of most economic instruments are appropriate standards, effective administrative, monitoring and enforcement capacities, institutional coordination and economic stability (Global Water Partnership, 2000). Designing appropriate economic instruments requires simultaneous considerations of efficiency, environmental sustainability, equity and other social concerns and the complementary institutional and regulatory framework. Some notable examples of these economic instruments include water prices, tariffs and subsidies, incentives, fees and fee structures, water markets and taxes.

Policy inferences

Institutional arrangements establish the interface between government and the private sectors in watershed and water management, with management typically involving a mix of government and private sector activity. Once this mix has been decided upon, the next step is selecting the policy instruments that will work best (see above). Some combination of policy actions and instruments is usually more effective than only a simple action or an instrument. A hypothetical example of such a combination is a rapidly growing city that faces the need for augmenting its water supplies to meet a growing population of people. Instead of choosing a costly option of augmenting water sources through water transfer arrangements or the construction of dam and reservoir systems, a responsible water manager decides on a strategy of replacing leaky pipes in the water-distribution system, charging higher water fees and/or providing users with water conservation knowledge and assistance.

A water policy will likely need to change incentive structures. However, policies and organizations can be changed to provide water managers with a strong incentive framework to improve the efficiency and equity of water distribution. This might be accomplished by providing the water users more responsibility for the costs and benefits of water deliveries and allocations. Another approach might be to assign water users with tradable water rights and then allow the users to employ water managers – much like what is done with some irrigation systems in Mexico. A third possibility could be having a water manager's salary largely dependent on the efficiency of water delivery and/or a percentage of fees collected in payment from water users. In several countries, such as the Philippines, water managers receive a bonus for good service or when a high percentage of farmers (90%) pay their water fees. The important point is forging a strong link between those using the water and those managing it.

Other incentive considerations

Incentives are also needed to encourage water users to make efficient use of decisions concerning their water supplies. However, this approach can be difficult to accomplish when many users are involved or when monitoring is difficult. The two most effective instruments in this case are water markets and prices that are based on the opportunity cost of providing the water. Water markets are probably the easiest means of introducing the incentives for efficient water use if the necessary water rights have been established and efficiently allocated. Where it is not possible to allocate water rights to users, water pricing – although not as flexible as water markets – can provide needed incentives. This option works best when water is piped directly to domestic and/or industrial users. It is much easier to meter the delivery of water flowing through a pipe than the delivery of water to widely dispersed farmers. To lessen the impact of higher water prices on low-income families, the price increases can be combined with technical assistance in implementing water conservation measures. Bogor, Indonesia, reduced water use by over 50% by using such measures as price increases and conservation measures.

Still another option for increasing efficiency in water use is to restrict water supplies and, as a consequence, provide incentive for water users to conserve water and adopt innovative technologies to make better use of the water. When farmers are also allowed to trade water, we obtain an added efficiency of moving more of the available water to the most productive farmers. Gleick *et al.* (2005) have suggested that California could meet all its future water requirements through conservation practice and instituting incentives to use water more efficiently.

A key factor to be remembered is that people need incentives to improve water use and allocation. Where these incentives have been effectively changed, major improvements in water use have occurred for both users and managers. About 30 years from now we may all be surprised by what has been achieved. Who could have predicted 30 years ago that we would have an international market for bottled water? Nevertheless, it exists and is a thriving source of economic gain.

Privatization of water

Privatization of water has become common throughout the world. As pointed out by Gleick *et al.* (2002), one of the most important and controversial trends in the global water arena is the accelerating transfer of the production, distribution and/or management of water and water services from public entities into private hands – a process loosely called *privatization*. Considering water as an economic good and privatizing water systems are not new ideas. Private entrepreneurs, investor-owned utilities or other marketing tools have long provided water or water services in different countries of the world. What is new today is the extent of privatization efforts under way and the growing public awareness of and attention to problems associated with these efforts.

There are many forms of privatization. In some cases, it is partial privatization with public–private partnerships being established. Only a certain part of the supply chain is privatized in other cases. In still other cases, as illustrated by the bottled water market, private water enterprises exist alongside the public counterparts – most drinking water is still not priced in the marketplace but comes through public waterworks. Elsewhere, large private water companies provide all of the water to local residents and industries.

Gleick *et al.* (2002) indicated that the revenues of such companies around the world are around $300 billion per year and that excludes revenues from bottled water.

Private water companies spring up because of the lack of ability of public water works to keep up with demand. Sengupta (2006) provides a graphic account of such a situation in New Delhi, India, as follows:

> It is a 'rare' morning when water trickles through the pipes. More often, not a drop will come. So, Mrs. Prasher will have to call a private water tanker, wait for it to show up, call again, wait some more, and worry about whether enough buckets are filled in the bathroom in case no water arrives....As the city's water supply passes through a 5,600 mile network of old and battered public pipes, 25 to 40 percent of the water leaks out. By the time the water reaches Mrs. Prasher, there is hardly enough. On average she gets no more than 13 gallons a month from the tap....That means that she has to look for other sources and scrimp and scavenge to meet her family's water needs. She buys an additional 265 gallons from private tankers for $20 a month. On top of that, she pays $2.50 to the worker who pipes water from a private tube-well she and other residents of her apartment block have installed.

While privatization of water is progressing at a fairly rapid rate, so is the opposition to privatization. Opposition is based on communities not wanting to lose control over what happens to one of their most necessary and important resources, their environmental concerns and a concern that some people will be excluded from obtaining adequate water because they cannot pay what the private companies are asking for water and water-related services (Wolff and Hallstein, 2005). Gleick *et al.* (2002) suggest that the greatest need for water services often exists in those countries with the weakest public sectors. However, at the same time, the greatest risks of failed privatization also exist where governments are weak.

A set of principles and standards for achieving water privatization is offered in Annex 3.2.

Water prices, subsidies and cost recovery

When water is considered an economic good, the objective should be full recovery of the cost of providing the water subject to constraints imposed by equity concerns. Importantly, all people should have access to water to survive regardless of economic ability to pay. This concept introduces the common practice of wealthy people subsidizing poorer people in terms of water supplies and full cost recovery. At a minimum, wealthy people and large industries should be paying the full cost of their water supplies. However, it is a *mixed situation* in the case of irrigation water for agricultural production, since many farmers cannot afford to pay the full cost of irrigation water because of the large infrastructure investments in many irrigation schemes. Subsidization can take place in the interest of food security and other welfare objectives in these cases. At the same time, appropriate pricing of irrigation water can provide farmers incentive to move to more efficient agricultural cropping patterns and use of their water.

Setting the fee structure and levels right is a key aspect of water pricing. Although flat-rate fees have been shown to result in inefficient and often overuse of water, incremental fees where there is a charge per unit of water, which increases with the amount of water used, have proven to be effective in many situations (Box 3.2). Water rationing can also help to provide incentive for increasing efficiency in use.

The same kind of reasoning applies to wastewater disposal, where one should apply the *polluter pays* principle from an economic perspective. That is, those disposing of wastewater should be covering the full cost of disposal in a manner that meets accepted environmental standards. Once again, some regulatory measures might also be needed to establish maximum levels of pollution that are permitted. The capacity of the environment to absorb waste is limited, and, therefore, the aim should be to have people use water in the most efficient and effective way possible, while minimizing the resulting waste. Also of relevance is the use of taxes, another economic instrument that can provide incentive for efficient and effective use of water.

Payment for environmental services

Of increasing interest is the practice of payments for environmental services (PES). Often in the past, the users of land, water and other natural resources were given what was called *subsidies* for certain environment-friendly actions on their part. However, recent thinking places such payments into a market-incentive mode and in the context of non-market prices. The payments are therefore looked at as legitimate PES rather than subsidies – a term that has negative connotations for many people. The extent of payment for environmental services is determined by applying non-market valuation approaches (see Chapter 4). Increasingly, practices and programmes are being developed where downstream land and water users pay upstream landowners and users for environmental services that maintain water quality or control downstream flows for the benefit of the downstream land and water users. In countries like Japan, such PES schemes have existed for many decades. Other countries like Colombia have tried the PES approach with mixed results. In Costa Rica, income generated by the national hydroelectric power company is taxed; these taxes are the basis for payment to upland watershed dwellers who undertake land-use practices that reduce sediment delivery to hydropower reservoirs.

Governmental Agencies, Land and Water User Groups and other Organizational Mechanisms

Many organizations and formal and informal groups affect the use and management of land and water resources within a watershed framework. Here, we focus on two groups – government agencies at all levels including quasi-governmental groups such as river basin commissions; and society-dominated water user groups, associations, partnerships and other private groups that might still have some local government input as one of the many partners.

Government agencies, boards and commissions

A large number of governmental agencies can be involved in the workings of a large watershed or river basin in terms of its functioning as a cohesive unit for land and water management. In the case of the USA, the federal government alone has 19 agencies that deal with water and watershed management practices (Table 3.1). Above it there are hundreds of state and local governmental agencies that are also involved in water and watershed management. Therefore, it is no wonder that conflicts arise between these agencies because of differences in mission for each one of them.

The Committee on Watershed Management of the US National Research Council provides a succinct overview of the situation. Organizational fragmentation is often a major obstacle to effective watershed and water management (CWM/NRC, 1999). To begin with, divisions among levels of government – federal, state and local – can generate genuine disputes over the proper locus of taxing, spending and/or regulatory authority. In addition, each governmental level might have different agencies pursuing apparent cross-purposes. One state agency can advocate a new dam while another might oppose it. One local agency might advocate locating a new sewer outfall at a certain place while another might oppose it. Such apparent conflicts among agencies are inevitable in a governmental structure that by design represents varied groups of stakeholders. Nevertheless, the varying levels of government are generally in pursuit of common goals. Certainly, those levels empowered to act can have some conflicts with their authorities, but these conflicts are far less significant than those that arise over how the land, water and other natural resources of a watershed might be used. For example, a fisheries management organization might view (correctly) a decision by a water and sewer authority to locate a sewer outfall near an oyster ground as having a negative effect on their goals of promoting oyster production and harvester's income. Governments must choose between legitimate but competing public purposes. Therefore, in this example, governments in general decide between the water and sewer authority's preference for locating a sewer outfall near an oyster ground, and the preferences of the fisheries organization.

Table 3.1. Federal agencies and their responsibilities in watershed and water management in the USA (1995). (From Committee on Watershed Management, 1999, adapted from O'Connor, 1995.)

Water supply	Flood risk management	Water quality	Erosion/sediment control	Ecological diversity/restoration	Flow regimes	Fisheries	Wildlife	Preservation	Recreation	Navigation	Hydropower	Research and dissertation data	Wetlands	Oceans and estuaries	Agencies
															Department of Agriculture
			●										○		Farm Services Agency
○	○	○	○			●	●	●	●			○	○		Forest Service
		○	●	○	○								●		Natural Resources Conservation Service
○		○	○												Agricultural Research Service
															Department of Commerce
		○				●	○		○				○	●	National Marine Fisheries Service
	●	●				●								●	National Oceanic & Atmospheric Administration
															Department of Defense
○	●	○	○	●	●					●	●	○	●		US Army Corps of Engineers
															Department of Energy
											●				Federal Energy Regulatory Commission
															Department of the Interior
		○	○	●	●		●		○			○	●		Bureau of Land Management
●	●	○	○			○	○		○		●	○			Bureau of Reclamation
		○				●	●	●	○			○	●		Fish and Wildlife Service
●	○	○	○	○	○	○	○					●	●	○	Geologic Survey
		○	○			●	●	●	●				○	○	National Park Service
	○	●	○			○	○						○		Bureau of Indian Affairs
															Department of State
●	●	●											○		International Boundary Commission
															Other Federal Units
○		●	○	●	○	○	○	○	○			●	●	○	Environmental Protection Agency
●	●	●	●	○	●	○	○		○	●	●	○	○		Tennessee Valley Authority
●	●	●	○	●	○	●	○			●	●		○		Bonnevilm Power Administration
	●														Federal Emergency Management Agency

Open circle indicates some related responsibilities; filled circle indicates significant responsibilities.

Within this often-conflicting administrative and managerial structure, decisions relative to allocating watershed and water resources among competing uses are made through a bargaining process among the same levels of government as well as vertical organizations (CWM/NRC, 1999). Policy for a specific action results from the formal and informal ways that organizations and their leaders seek to influence each other by economic assessments, environmental impact statements, as well as water quality measurements, recognition of policy constraints, exchanges of support and exchanges of both threats and promises.

In countries with federal systems of government, each state, province, canton or other political designation has its own organizational structure for dealing with watershed and water issues. For example, a natural resources or environmental protection agency might be responsible for water supply and quality, another agency can deal with water-based recreation, a wildlife and fisheries agency is responsible for aquatic life and an economic development or public works agency regulates dam construction and navigation. Some states, provinces or cantons within a country bordering an ocean or sea can also have coastal zone management organizations or commissions to deal with land-use issues on coastal watersheds and estuaries. Others might have *special wetlands boards* to oversee wetlands protection and use. There can be regional planning commissions or districts that deal with issues and proposals that go beyond one third-tier jurisdiction's boundaries. Some units of government might join together with federal government agencies to form river basin commissions. Cutting across all of these local, state and federal efforts there can be federal regulatory agencies dealing with interstate water transfers, fish and wildlife resources, as well as other natural resource and commerce issues that transcend local government jurisdictions.

The idea of interagency committees for major river basins started in the USA in the 1940s and grew in the following years to be eventually replaced by more formal river basin commissions in the 1960s and 1970s (Featherstone, 1996). However, these commissions have not necessarily fared well. The responsibilities expected of watershed organizations replicated some of those already existing within federal and some state agencies. There was an understandable reluctance on the part of federal and state agencies to transfer the authority they held to empowering these organizations. Without the needed authority, the large-scale river basin commissions could not become effective mechanisms for river basin coordination and planning.

As mentioned earlier, watershed and water management coordination and integrated planning are not introduced into a vacuum. There exists at any time a complex set of policies and relationships between agencies, groups of stakeholders and others in the private sector. To fit a newly formed regional organization into such a complex set of existing institutional arrangements is difficult at best. It is bound to disturb the existing and, in some cases, fragile balance of power that exists. Established political alliances need to be convinced that the large-scale river basin entity can be desirable and will produce superior results in the context of their vested interests (CWM/NRC, 1999). A few of the larger-scale river basin commissions still exist, but there has been a shift in focus in the USA toward water quality as well as quantity issues. The Environmental Protection Agency (USA), for example, did not support the larger river basin commission approach.

As mentioned earlier, management of land, water and other natural resources has evolved and the support for a watershed approach to planning and management is stronger than ever. But, the days of mammoth river basin commissions are not returning,

> **Box 3.3. Partnerships in watershed and water management: two examples.**
>
> Even though the federally operated Tennessee Valley Authority (TVA) in the south-eastern USA has adopted a watershed approach, there has been a steady move over the years to a stronger partnership approach, involving private landowners, soil and water conservation districts, local government, as well as state natural resource and fish and wildlife agencies in the project. TVA finds itself working under varying local land-use laws, wildlife laws and other approaches to water-related objectives. TVA – even as a federal agency – has had to rely on the powers and persuasion of local entities and the states to accomplish its water quality goals (CWM/NRC, 1999).
>
> New Zealand is an interesting case in terms of watershed focused planning both because of the long history (since 1968) of using watersheds as administrative units and because local governments reorganized and consolidated in 1989 to create 16 new regions defined by watershed boundaries (Dixon and Wrathall [1990] as cited in CWM/NRC [1999]). Because many of the new established watershed boundaries were the same as those used by the New Zealand River Boards created in 1986 to deal with flooding and erosion problems, the newly recognized watershed units were familiar to people and, therefore, accepted by citizens and administrators.

although there are some successful examples still in existence. Rather, there seems to be an emphasis on partnerships between governmental bodies, between government and water and land users, and between environmental groups and private industry (Box 3.3).

The World Bank has recently released a summary study of research on the effectiveness of decentralized partnerships for river basin management in different countries of the world. Some of the main findings and lessons learned, both from the eight case study countries and from a global survey of some 82 decentralized river basin organizations (RBOs) include the following (Kemper *et al.*, 2006):

> While the level of economic development of the nation and the basin can clearly make the creation and sustainability of basin-level institutions more or less difficult, there is no reason to believe that improvements in water resource management are limited to wealthy basins. Notable improvements can be and have been realized in a variety of settings, and sometimes very early in the life of basin organizations and stakeholder participation initiatives.

> Although river basins are important hydrologically, ecologically, and economically, not all aspects of stakeholder participation and not all decisions and activities that contribute to IWRM need to be organized at the basin scale. The "lowest appropriate level" for some water-resource management functions may be a sub-basin, a local or regional unit of government, or a hybrid unit sometimes referred to as the "social basin", (e.g. the basin subcommittees in the Alto Tiete case).

> The establishment of participatory and decision making structures involves shifts of power, something which can be a controversial and complicating factor. Even in settings where there is a desire for decentralized basin management, the political dimensions of public policy play a key role. Efforts by the Andalusian regional government to exercise more leadership over basin management in the Guadalquivir case, the efforts of users in the Jaguaribe basin to gain more influence in decision making, especially with regard to infrastructure, the desire of the state company in Indonesia to also take on pollution control which is currently the responsibility of provincial government, are only a few

examples of the ways in which jurisdictional and other power-related considerations are likely to arise.

Finally, decentralization reforms and the establishment of river basin management with active stakeholder involvement are processes that take time, sometimes even decades. In order to sustain the reform process, consistent support is vital, as is the ability to adapt and modify basin management arrangements in response to changed conditions. Central governments and external organizations promoting integrated water resources management on a river basin scale must be prepared to sustain their commitment to reform, both through changes of administration and over the long haul.

Water user groups and associations

One of the more interesting recent developments is the growth of small community-based watershed and water management groups and associations of various kinds. Most of them have the purpose of helping to make management and use of watersheds and water resources as effective and efficient as possible and to meet the needs of stakeholders. The alliances that these groups – often a combination of local government and society groups – bring to the table are formed by necessity or because of a set of problems or opportunities that emerge. One example of this type of quasi-formal alliance rallying around a set of local issues and opportunities was recently described in the *Oregonian*, a widely distributed newspaper serving the people of Oregon (Box 3.4).

Governments have long been encouraging these types of alliances, recognizing that public funds to implement watershed and water management are scarce (Hodgson, 2003). There were more than 1500 locally led watershed and water management initiatives in the USA by 1999; almost all were established since 1990 (Lant, 1999). Participatory or co-management of natural resources is therefore a growing phenomenon worldwide. (We use the term *co-management* to refer to schemes that involve both government agencies and society groups such as communities, cooperatives and associations.) Although we have not found an inventory of local watershed and water management initiatives in 2006, it is safe to say that they number in the many thousands in the USA alone. Similar groups have been formed in European countries and they are also springing up in other regions of the world.

Some of these groups come together for action because much of the information that surfaces creates an almost crisis situation, i.e. the coming together of people due to a push. An example from the city of Philadelphia in the eastern USA is illustrated by a newspaper article in the *Philadelphia Inquirer* (see Box 3.5). This example is also an interesting illustration of two other points – the need for good information in order to plan for water resource management and the bureaucratic nature of government rules and regulations.

The types of groups that currently exist in the USA and elsewhere have assumed different titles and have different scopes depending on their purposes. At a broad level, there are such groups as *watershed councils*, while at a more narrow level, there are water user organizations that are generally focused more explicitly on allocation and security of water for specific uses such as irrigation. Watershed councils are formally or informally established groups oriented to local issues and function as non-profit advisory, educational and/or advocacy organizations that encourage the protection, conservation and sustainability of water and watershed resources on a river basin basis. Some watershed councils act as fora for exchanging ideas, views, concerns and recommendations, while others are actively responsible for the management of a river basin. Membership in these councils is typically open to public agency personnel and private-sector stakeholders with vested technical and sociocultural interests in a river basin and its tributary watersheds. The widespread nature of the interest in watershed councils is illustrated by the nearly 8 million entries that emerge when searching the Internet for them. Watershed councils in the USA and other countries reflect the issues of crucial concern to the watershed managers and stakeholders in the river basin and geographic region of interest.

Box 3.5. New maps put towns in tricky position and force choices. (From Mastrull, 2006.)

A choice had to be made between using a new set of flood plain maps done by Temple University or the previous outdated, imprecise maps provided for three decades by the Federal Emergency Management Agency (FEMA). The sets provided different information about flood plain use; and the implications are serious for the towns involved. Mastrull writes:

'The borough is now in a dilemma as to what map we can adopt', said Jeff Heller, zoning officer for Hatboro, where Temple researchers found 19% more homes and businesses at high risk of flooding than FEMA's map shows. 'Do we put our municipality at risk by not following the Temple map?'

The Temple maps – the subject last week of the Inquirer series 'A Flood of Trouble' – were formally presented Friday to 30 town managers, zoning officials, engineers and planners from the 56-square-mile watershed, a sodden sector containing 11 municipalities with 300,000 people and stretching from Bucks County through Montgomery County and into north-east Philadelphia. Since 1999, 14 flood-related deaths have occurred there, and in the last three decades nearly $30 million in claims have been paid out by the FEMA-run National Flood Insurance Programme.

Under a rising threat from stormwater runoff and creek overflow, watershed municipalities contributed a total of $70,000 to Temple's $700,000 remapping project.

FEMA, likewise, was eager for more accurate maps of the Pennypack. With $3 billion in recharting needed nationwide to update its aged archive of 100,000 charts, FEMA had been looking for mapping partners. When Temple offered four years ago to study the Pennypack, FEMA produced a $192,500 grant.

No one anticipated that FEMA would reject the maps, and no one would have guessed the reason: Temple's precision exceeded the agency's standards.

The Temple team found 3.4 square miles in the watershed at high risk of flooding – 24% more than shown on FEMA maps. Temple also counted 708 mostly residential buildings in that vulnerable land, an increase of 131.

Those findings, however, were based on a level of detail that violated the agency's uniform standards for mapping, FEMA officials informed the Temple researchers late in the summer. The researchers had surpassed those standards by including small tributaries typically discounted by FEMA, and by including clogged culverts and storm drains in their flood-risk calculations.

Unless amended, Temple's maps will not be added to the federal map collection, which is the foundation of the National Flood Insurance Program. Communities must recognize the FEMA maps if their residents are to be eligible for flood coverage, and mortgage lenders are expected to use them to determine whether borrowers need the policies.

Ken Wallace, an official from FEMA's Philadelphia headquarters, attended the meeting Friday to try to explain. 'We are not accepting [the Temple maps] as they are presented now', he told the perplexed crowd of not only municipal officials but representatives of environmental groups and aides to state and federal legislators.

Temple has been asked to recalculate the Pennypack's flood plains based on 'our mapping standards for what essentially is just an insurance program', Wallace said. However, he conceded that the Temple maps might provide more reliable flood-risk information than FEMA's.

Continued

Box 3.5. *Continued*

'If you feel they better serve your community, please adopt them and use them for flood plain mapping purposes', he said.

Jeffrey Featherstone, director of Temple's Center for Sustainable Communities and head of the Pennypack project, also urged the municipal officials to adopt the maps and use them in crafting land-use ordinances.

Box 3.6. Watershed Districts in the State of Minnesota, USA (www.mnwatershed.org/): an example of locally based organizations established to achieve integrated watershed management.

Minnesota has formed an Association of Watershed Districts that informs citizens about their watershed districts and their role as special purpose local units of government that work to solve water-related problems. As of 2006, there were 46 watershed districts in the state, ranging in size of from 112 to 15,518 km², with boundaries that coincide with natural watersheds, not political units. Watershed districts are established when water management problems extend beyond the capabilities or jurisdiction of a single community or city; they can be formed by local residents, cities or county boards by petitioning a state agency, called the Board of Soil and Water Resources. The underlying premise is that managing natural resources on a watershed basis makes good sense and favours a more holistic approach to resource conservation. Each watershed district is governed by a board of managers, appointed by the boards of commissioners of the counties that have land within the watershed.

Within their respective watersheds, 'Watershed Districts in Minnesota . . .

- Are partners in water planning and management with state, counties, cities and soil and water conservation districts;
- Are partners in wetlands protection and management. . . .;
- Conduct water quality surveys of lakes and streams within the district;
- Monitor groundwater levels;
- Manage draining systems;
- Regulate, conserve and control the use of water within the district;
- Provide for wildlife and enhance recreational opportunities as benefit of projects to improve water quality and provide flood protection;
- Establish, record and maintain hydrological data;
- Approve culvert size and placement in all roads of the district; and
- Other projects related to meeting the purposes of the district.'

Other water user organizations have been in existence for centuries. They are particularly popular for such organized activities as irrigation systems, but they also exist for other forms of watershed and water use. Oftentimes, the main members of a water user organization are farmer associations and individual farmers. Many of them have been formally established in a legal sense and have been afforded certain authorities from the government via statutory mechanisms (see Box 3.6).

Successful water user organizations are those with a clear agenda that is independent of political situations, that possess identifiable compliance responsibilities among the multiple parties and have obtained that funding necessary to implement the agreed to programmes (Toupal and Johnson, 1998; Johnson, M.D. 2000). A lack of thrust that causes private interest groups, local citizens and NGOs to assume an adversarial role concerning the actions of public management and regulatory agencies must not be allowed to occur. It is also important that all parties in the organization have access to all of the technical information necessary to make decisions relative to the management of water and watershed resources. Effective agendas for these organizations include non-commodity aspects of watershed-based resources such as recreational opportunities, landscape beauty and indigenous beliefs about land, water and natural resources. The increasing demand by people to incorporate the *multiple-use concept* management planning is crucial. No longer is management driven by only the interests of a single group of stakeholders or one economic concern.

4 Planning and Policy Making

Planning is a term given to whatever orderly process we use to understand and deal with future trade-offs, uncertainties and complexities that hinder clear, non-controversial decisions and actions. Planning becomes necessary because there are trade-offs between different interests and priorities among the various legitimate stakeholder groups on a watershed and there are less easily defined intergenerational trade-offs – what we do today on a watershed or river basin affects the options available to future generations. In a practical sense, there are trade-offs between cost of doing something versus quality and quantity of output achieved. The constraints include political, policy, economic, social or other institutional and technical ones. Effective IWM involves many trade-offs, constraints and complexities; so, as indicated in Chapter 3, planning, including conflict management and resolution, is considered essential in order to achieve the most effective, efficient and equitable results possible.

The first part of this chapter presents the general context for watershed management planning. This discussion is followed by a discussion of background on the move away from planning dominated by political boundaries to planning based on watershed boundaries and that is participatory involving varying governmental and non-governmental stakeholders. The lessons learned from experiences to date with coordinated and participatory watershed planning are discussed in the third section; and the final section discusses some of the overall approaches to planning and the economic assessment methods used within these approaches to make sense of the trade-offs between diverse activities and outcomes. Economics provides a unique opportunity to compare most events and outputs using a common metric. Annexes to the book contain suggestions where one can go to obtain the technical details.

Setting the Context for Planning

As implied in Chapter 3, several sets of linkages and trade-offs have to be considered in developing a framework for planning watershed management and use. Therefore, we see that we have:

- Spatial linkages and trade-offs in relation to political boundaries and potential transboundary water-related conflicts that might exist on a given large river basin or watershed and the impacts downstream from actions upstream. The key here is to plan in such a way so as to achieve an equitable and acceptable balance between the interests of upstream and downstream stakeholders and between the different political units within a watershed or river basin.
- Interest-group trade-offs within a given jurisdiction, upstream or downstream area. The fact is that in any given area – even small townships – there are multiple and

diverse legitimate stakeholder interests in the watershed, ranging from pure economic production objectives to pure protection and environmental objectives. Conflicts between interests are almost bound to exist in any given area. While we try to plan for the win-win situations, we most often end up with trade-offs between the different sets of values involved. A land owner or land user can support the environmental protection of land situated above the person's own parcel of land while equally supporting their right to produce to the maximum on their land regardless of the ultimate consequences. This dual view of land and the issue of environmental vis-à-vis production objectives can create issues in terms of broader watershed-level planning exercises. The challenge in planning for IWM is to find the mix of trade-offs that is fair and generally acceptable to the stakeholders. This involves planning for legal measures, incentives and other mechanisms for generating a consensus view. It involves planning for both 'carrots' and 'sticks' or both incentives and enforceable regulations.

It is also necessary to consider the following two sets of trade-offs from a planning perspective:

1. Temporal trade-offs – i.e. trade-offs between the interests of current and future populations – the so-called *intergenerational* trade-offs. What the current population does on a watershed can markedly affect what happens to the next generation or a distant future population, and – as discussed in Chapter 2 – there can be cumulative effects over time that lead to a future disaster such as when the water table is steadily drawn down because of people's non-sustainable use until the time when the water is gone and disruption of life occurs as a consequence. The key here from a planning perspective is to balance the interests of the present set of stakeholders with the opportunities that are left for future generations to achieve what they aim to achieve. In some cases, there can be win-win situations, where what people prefer to do today in fact turns out to be best for future populations. However, this favourable situation does not exist in other cases. Also, the key is not to try to rigidly define what future populations will need, but rather to expand the number and breadth of options that future populations will face. This is central to the concept of *sustainability* that has evolved over the past couple of decades.
2. Cost and outcome trade-offs, that is, the fact that in most situations, greater expenditure of resources and effort leads to better solutions and outcomes up to some point. This occurrence in the real world of scarce resources is a trade-off in planning that is always faced. We would spend as much as is needed to reach an optimum in the ideal world. In the real world, however, we are told to optimize results given a resource constraint. This is also related to the role of research, technology development and use of all the new information transfer and communication technologies (see Chapter 7). While watershed planning in the real world – as distinct from in theory – is mostly political in nature, good planning should also involve technical input related to knowledge of hydrology, planning methods, economics, land management and means to reduce risk to acceptable levels.

Regarding this latter set of trade-offs, although this chapter focuses mostly on approaches and institutional issues related to planning, it needs to be remembered that the activities of successful planning for IWM depend on a thorough and broad understanding of the biophysical and engineering relationships involved in good watershed management and also knowledge of what the impacts are of different human interventions. That is why this book emphasizes both sets of relationships – i.e. biophysical and institutional. In the final analysis, despite the heavy hand of politics in most of the real-world watershed management plans, knowledge of such technical relationships are central determinants of what a good plan needs to contain.

A couple of other contextual points should be in mind when a group of stakeholders implements a process of planning for IWM. These are that:

- There is always an existing policy framework and most often some kind of plan or set of plans exist and affect the management and use of the watershed. Therefore, planning does not start in a vacuum. There is always a present situation to consider – one that has often resulted from previous planning, formal or informal, and implementation efforts of various sorts. Every watershed has a history of institutional development and policies in place.
- In most cases, the proposed planning effort is part of a dynamic, ongoing process of adaptation as new knowledge emerges and as the balance of interests among stakeholders shifts. With the new technology transfer mechanisms, the evolution often ends up being more dynamic and rapid since better information is communicated in improved ways. Making a detailed plan for 10 or 15 years and expecting it to remain intact and be followed over that period of time is naive. Many adjustments and revisions will likely be needed and will take place as new knowledge emerges and as interests change. Planning, like sustainability, should be looked at as a dynamic process and not an end state. An IWM plan is a living document.

These two points relate directly to the types of planning processes that are needed and the type of plan being developed. Normally, we think of three types of plans:

1. A *strategic plan* provides an agreed upon – or consensus – vision of how the watershed should evolve in the future and the strategic elements needed to move towards that vision. A strategic planning exercise often starts with an internal assessment of strengths, weaknesses, opportunities and threats before going to the public and other stakeholders for input. Strategic plans need to be participatory if they are to have any meaning in a political context.

2. A *tactical plan* – often called a medium-term plan – lays out more specifically and concretely what is needed in the next few years and the milestones for monitoring progress with intended actions often laid out in a logical framework format.

3. An *operational budget plan* – often called the programme of work and budget – for the coming year or biennium with specific assignments, expenditures and actions specified in greater detail.

In recognition of the dynamics of the process and uncertainties that are sure to surface with new information and unexpected events, the tactical plan sometimes is undertaken annually but on a rolling 3-year basis; that is, the 2006–2008 tactical plan produced at the end of 2005 is revised at the end of 2006 so the next plan would be the 2007–2009 tactical plan with years 2007 and 2008 revised from the previous tactical plan. This process allows continuity with all the dynamic elements entering the picture when various groups introduce changes that need to be made in the 2007 and 2008 plans. At the same time, it keeps all of the stakeholders aware of what is expected over 3 years, given the best knowledge available at that time. This annual process of updating continues year to year.

The emphasis here is on development of longer-term strategic watershed management plans and the process of interaction involved in developing an overall strategy and plan for managing a watershed or river basin. Many stakeholder organizations and groups generally participate in the overall strategic planning exercise; and many of them will have their own distinct forms of tactical plans and programme of work and budgets that have to fit within and, in turn, influence the context of the overall strategic and participatory watershed management plan (see stakeholder categories in Annex 4.1).

Watershed-level Planning and Action

Because political and tenure boundaries rarely coincide with natural watershed boundaries, political institutions have not considered that watersheds are workable land management units for planning and action in the past. After all, the actions have to be carried out by individual government units, NGOs and private entities that often cross watershed boundaries. This viewpoint is rapidly changing as illustrated in the earlier chapters. Most countries are moving towards systems of IWM planning or even larger river basin-level planning, as mentioned, the terms differ and the complexity varies, but the basic principles remain the same. In most developed countries and many developing countries, it is being recognized that planning for IWM makes sense from an environmental and economic point of view that takes sustainability of economic activity and trade-offs into account.

It was not until the Dublin Conference on Water and the Environment in 1992 and the UN Conference on Environment and Development in Rio de Janeiro, Brazil (1992), that a more comprehensive approach to water management using watershed boundaries as a basis was judged by key high-level policy makers to be necessary for sustainable development. This awareness together with the need for participatory institutional mechanisms related to water called for a new facilitating and coordinating organization. In response to this demand, the World Bank, the United Nations Development Programme (UNDP) and the Swedish International Development Agency created the Global Water Partnership (GWP) among nations in 1996. The first of six objectives for the GWP addressed the need for IWM as follows:

1. Support-integrated water resources management (IWRM) programmes by collaboration – at their request – with governments and existing networks and by forging new collaborative arrangements.
2. Encourage governments, aid agencies and other stakeholders to adopt consistent and mutually complementary policies and programmes.
3. Build mechanisms for sharing information and experiences.
4. Develop innovative and effective solutions to problems common to IWRM.
5. Suggest practical policies and good practices based on those solutions.
6. Help match needs to available resources.

The GWP initiative was focused on promoting and implementing an integrated approach to water resources management through the development of a worldwide network that could pull together financial, technical, policy and human resources to address the critical issues of sustainable water management and provide shared experiences among countries (Global Water Partnership, 2000). The GWP is a working partnership among all those involved in water management: government agencies, public institutions, private companies, professional organizations, multilateral development agencies and others committed to the Dublin-Rio principles. This comprehensive partnership actively identifies critical knowledge needs at global, regional and national levels, helps design programmes for meeting these needs and serves as a mechanism for alliance building and information exchange on IWRM. As mentioned earlier, this is duplicated by thousands of such integrated partnerships at state and local levels, partnerships that act as clearing houses and coordination units for the planning, monitoring and evaluation of activities on all from small, local watersheds up to large, multinational river basins.

Other international groups have also been promoting integrated approaches to planning and managing river basins. For example, to help coordinate action for living rivers worldwide, the World Conservation Union (IUCN) and leading global water and environmental organizations have launched a network of experts, practitioners, policy makers, local community representatives and end users at the World Water Week on 20 August 2006. The network is focused on supplying Environmental Flows approaches to leave enough water in rivers to maintain downstream benefits for people and nature. Environmental flows mean that enough water is left in streams and rivers and they are managed to ensure downstream environmental, social and economic benefits. It includes planned releases of water from dams and other infrastructure. Releases of a minimal amount of water are alternated with larger amounts to cause rising water levels in the river and limited floods downstream. The goal is to maintain the stream or river in a healthy state as agreed between the many land and water users in the basin. The initiative has a 5-year action plan for the 80 partner organizations to improve water management for healthy rivers and healthy communities. Demonstration of good management in seven basins is supported by the development of tools for financing, governance, empowerment and information.

The main goal of the IUCN Water and Nature Initiative is the mainstreaming of an ecosystem approach into catchment policies, planning and management (available at: http://www.waterandnature.org). The interest in IWRM is not just an academic one and a matter of words and advice. For example, under the Asian Development Bank's Water Financing Programme for the period 2006–2010, investments in water are expected to double to more than $2 billion and be directed towards reforms and capacity development programmes, at rural communities, cities and river basins. As part of this initiative, the Asian Development Bank is targeting 25 river basins in Asia for the initial phase. Of interest in terms of the present discussion is the checklist of some 25 elements that the bank has put forth as a guide to what actions need to be dealt with in introducing, planning and operating IWRM efforts at the river basin level. The checklist and a brief description of the elements are shown in Box 4.1. Although focused on larger basins, the Asian Development Bank list provides a set of elements that also need to be considered in smaller basins and watersheds.

As discussed in Chapter 3, implementation of IWM requires as a prerequisite the introduction and stabilization of certain enabling conditions, at the national, regional and local levels, and perhaps at the international level. Therefore, the Asian Development Bank suggests that IWM is a long-term process that needs sustained commitment by all stakeholders in the river basin (see http://www.adb.org/water/Operations/WFP/basin.asp#2). Its implementation will take decades of incremental improvements to achieve full results. Introducing IWRM in a river basin needs a positive enabling environment, clear institutional roles and practical management

Box 4.1. Twenty-five elements in integrated water resources management: an Asian Development Bank checklist. (From http://www.adb.org/Water/Operations/WFP/basin-elements.asp)

The following 25 elements are widely accepted to be important in introducing IWRM in river basins.

Incorporating them into institutional reforms, development strategies and investment projects will make a significant difference for IWRM in the basin. Improvements may also be needed in the enabling environment at the national level.

	Typical interventions/criteria
1. **River basin organization**	Build capacity in new or existing RBOs, focusing on the four dimensions of performance (stakeholders, internal business processes, learning and growth and finance) under the Network of Asian River Basin Organization's (NARBO) benchmarking service
2. **Stakeholder participation**	Institutionalize stakeholder participation in the river basin planning and management process including active participation of local governments, civil society organizations (academe, NGOs, parliamentarians, media) and the private sector, and an enabling framework for meaningful stakeholder participation in project specific planning decisions
3. **River basin planning**	Prepare or update a comprehensive river basin plan or strategy, with participation and ownership of basin stakeholders and application of IWRM principles in land-use planning processes
4. **Public awareness**	Introduce or expand public awareness programmes for IWRM in collaboration with civil society organizations and the media
5. **Water allocation**	Reduce water allocation conflicts among uses and geographical areas in the basin with participatory and negotiated approaches, incorporating indigenous knowledge and practices
6. **Water rights**	Introduce effective water rights or entitlements administration that respects traditional or customary water use rights of local communities and farmers and farmer organizations
7. **Wastewater permits**	Introduce or improve wastewater discharge permits and effluent charges to implement the polluter pays principle
8. **IWRM financing**	Institutionalize models whereby all levels of government contribute budget to IWRM in the basin
9. **Economic instruments**	Introduce raw water pricing and/or other economic instruments to share in IWRM costs, stimulate water demand management and conservation, protect the environment and pay for environmental services

Continued

Box 4.1. *Continued*

	Typical interventions/criteria
10. **Regulations**	Support the development and implementation of a legal and regulatory framework to implement the principles of IWRM and its financing in the basin, including tariffs, charges, quality standards and delivery mechanisms for water services
11. **Infrastructure for multiple benefits**	Develop and/or manage water resources infrastructure to provide multiple benefits (such as hydropower, water supply, irrigation, flood management, salinity intrusion and ecosystems maintenance)
12. **Private sector contribution**	Introduce or increase private sector participation in IWRM through corporate social responsibility (CSR)-type contributions
13. **Water education**	Introduce IWRM into school programmes to increase water knowledge and develop leadership among the youth, including responsibility for water monitoring in local water bodies
14. **Watershed management**	Invest to protect and rehabilitate upper watersheds in collaboration with local communities and civil society organizations
15. **Environmental flows**	Introduce a policy and implementation framework for introducing environmental flows and demonstrate its application
16. **Disaster management**	Investments in combined structural and non-structural interventions to reduce vulnerability against floods, droughts, chemical spills and other disasters in the basin
17. **Flood forecasting**	Introduce or strengthen effective flood forecasting and warning systems
18. **Flood damage rehabilitation**	Investments in the rehabilitation of infrastructure after floods
19. **Water quality monitoring**	Initiate or strengthen basin-wide water quality monitoring and application of standards
20. **Water quality improvement**	Invest in structural and non-structural interventions that reduce point and non-point water pollution
21. **Wetland conservation**	Invest to conserve and improve wetlands as integral part of the river basin ecosystems
22. **Fisheries**	Introduce measures to protect and improve fisheries in the river
23. **Groundwater management**	Institutionalize and strengthen sustainable groundwater management as part of IWRM
24. **Water conservation**	Institutionalize a policy and implementation framework to promote efficiency of water use, conservation, and recycling
25. **Decision support information**	Improve online publicly available river basin information systems to support IWRM policy, planning and decision making, including dissemination of 'tool boxes' and good practices

instruments. The process can be anchored and its achievements monitored through a river basin organization (RBO) that includes:

1. Institutionalized stakeholder participation.
2. Comprehensive river basin planning and monitoring.
3. The enabling environment includes an effective water policy, updated legislation and conducive financing and incentive structures at the national level. Among the issues to be addressed through the enabling environment are the following:

- Cost sharing and recovery;
- Water-use rights;
- Responsibilities of national water apex body, RBOs, local governments, service providers and water user organizations and the private sector.

People are now recognizing that it is possible to overcome barriers to the adoption of policies that support an IWM approach across political boundaries (Brooks *et al.*, 1992). An increasing number of policy makers and major development banks such as the World Bank and Asian Development Bank are recognizing that environmentally sound watershed practices, projects and programmes must allow for the interrelationships between land and water uses on watersheds to achieve conservation and sustainable use of these natural resources. They realize that ignoring the boundaries and interrelationships set by the forces of nature will repeat the environmental problems of the past. Planning must take place in that context.

The recognition of the need for integrated watershed planning might come more rapidly in countries with unitary forms of government – since the central government is all-powerful and can dictate decisions that make planning across administrative boundaries mandatory. In countries with federal systems of government, where political boundaries of subunits are constitutionally set and where the central government cannot dictate to constitutionally established subunit governments, the process is more complicated but as important to implement. It can take a long time to get an effective IWM plan and its implementation running effectively. The case of Brazil, which is considered a model in some areas, was mentioned in Chapter 3. Even in Brazil, however, which is a federal country, it took 10 years to fashion an effective system (UNDP, 2006). In countries with democratic systems of governance, extensive negotiation and give and take is a necessary part of the process since, ultimately, civil society must be satisfied by consensus with the results. A classic example of this process of give and take between powerful cities and the individuals who live on the watersheds apply to New York City (National Research Council, 2000).

It is not possible to describe a general structure for government and non-governmental alliances for dealing with watershed resources that would apply in all countries with democratic systems of governance in this chapter. Each country has its peculiar set of government agencies, levels of government and types of civil society groups that deal with decisions about watersheds. The basic government units in Switzerland are Cantons. In the USA, it is states and counties within states, whereas it is provinces in Canada. However, it is possible to say in general how the process of decision making takes place regardless of whether they are unitary or federal systems of democratic government. A bargaining process emerges among agencies at any given level of government, among and between levels of government, civil society groups and the private sector and other stakeholder groups. This give and take – or bargaining for trade-offs that reflect different stakeholder priorities – is both formal and informal in the ways that leaders of organizations try to influence each other (CWM/NRC, 1999).

Influencing this process at various points are economic arguments and studies by various groups that are done on the impacts of proposed changes, environmental impact studies, technical studies of various kinds and studies that deal with policy and other constraints. The ultimate results include decisions on whether to change land and water management and uses of such resources. If it is decided that change is needed and is acceptable, then new laws are passed, new policies are designed and implemented, new sources of funding may be sought and possibly new agencies or subagencies are created. It is generally a dynamic process with possibly a distinct starting point but no fixed ending point. Rather, it mostly involves an iterative process of successive approximations as the overall planning process moves towards a consensus. Since the balance of interests and powers are constantly changing – for example, with each new election, with changes in economic activity, and with new environmental concerns emerging – so will the strategic watershed management planning exercise and the consensus reached. Therefore, it is a dynamic process as indicated in Fig. 4.1 and also discussed in Annex 3.1.

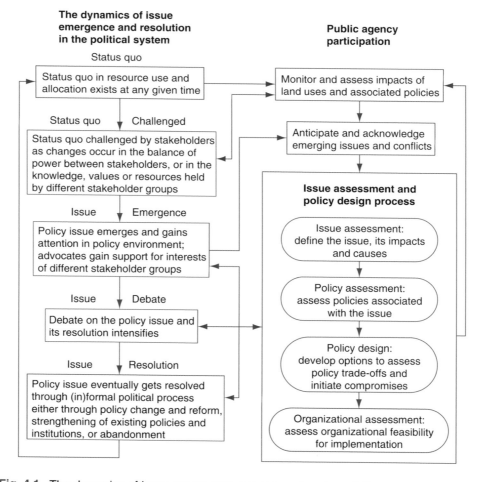

Fig. 4.1. The dynamics of issue emergence, assessment and resolution.

The point that is increasingly being recognized and acted upon is that some formal planning process is generally desirable. Some coordinating and integrating mechanism by which the various groups involved in decisions and in use of the watershed and those involved in doing the studies that are needed can come together in an orderly, positive and efficient fashion to organize and implement the issue resolution and bargaining process that is bound to take place because varying interests and priorities of different stakeholder groups are bound to exist. We have mentioned the river basin commissions and watershed associations as examples of such groups (see Chapter 3). The effectiveness of such groups generally depends on persuasion rather than law, which is a distinct advantage in countries where there is strong motivation for cooperation and social equity.

These planning groups might involve all stakeholders on a given watershed. However, in many cases, the process starts with subgroups – either ones with the same general interests come together to form a consolidated plan that they could support – or groups of people on subwatersheds come together for a common purpose of deciding on what is best for their entire watershed given all the interests and constraints that they face. The IUCN Water and Nature Initiative builds on this necessary process of coordination and integration of interests, as do many states in the USA.

There is an emerging trend towards decentralized but coordinated responsibilities for the environment and towards participatory planning and management of water resources and associated watersheds. The evidence of success of participatory approaches to sustainable development and ecosystem management is mounting. The way to the future will likely involve further development of innovative institutional mechanisms involving local participation and adaptive management with recognition that planning is a continuous process of evolution and successive approximations rather than a distinct, time-bound process involving definable starting and ending points.

The general elements in the planning process for IWM are shown in Fig. 4.2. The two main inputs to this process are technical – hydrology, ecology, economics, sociology and so forth – on the one hand and political negotiation on the other. If the two sets of inputs interact effectively in the planning process with each respecting the other, then the result hopefully is an informed and fair set of planning guidelines on the actions needed by all key stakeholders to move ahead towards a common set of goals.

Lessons from Past Experience: Lessons for the Future

A great deal of experience has accumulated with regard to what works and what does not work as coalitions of groups of stakeholders move ahead with watershed planning and the implementation of resulting plans. Some of this experience has been systematically gathered and brought together in various sets of lessons. A prime example is the work that the Environmental Protection Agency in the USA undertook a few years ago to bring together experience gained in the country (EPA, 1997). These lessons were originally developed by a focus group consisting of representatives from the River Network, Know Your Watershed, Center for Watershed Protection, the Maryland Office of Planning, several EPA Regional Offices and others and then

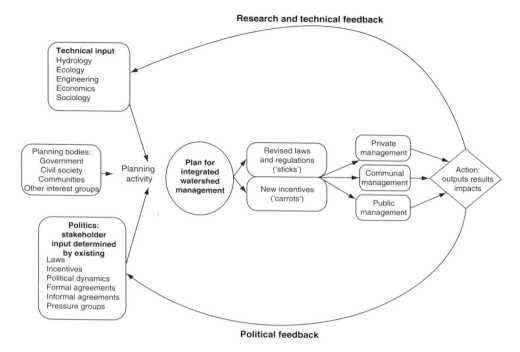

Fig. 4.2. The dynamic integrated watershed management planning and action process.

circulated to approximately 100 watershed coordinators and their supports. The EPA in partnership with other groups has been pursuing an integrated watershed approach to managing and protecting lakes, rivers, wetlands, estuaries and streams. There has been much trial and error in the effort and important lessons have been learned. Other groups that have been working with integrated watershed river basin management have also drawn lessons from their work.

In synthesizing from a number of these various sources, as well as the EPA lessons discussed above (EPA, 1997; Jones, 2001; McNally and Tognetti, 2002; World Wildlife Fund, 2003; Environment Agency, 2006), the following lessons emerge regarding effective planning for IWM ranging from the more or less formal planning for – and management of – smaller watersheds to the more formal and often international treaty-based planning and management for large river basins of the world. These lessons include:

- A consensus long-term vision needs to be established and exist for the watershed or basin and this vision needs to be clearly stated and effectively communicated. Productive planning depends largely on the existence of a clear consensus vision of where we want to go.
- Good communication and transparency are key elements in the planning and management process – this includes the need to have a common data and information base and set of definitions on which to base communication.
- Integration at all levels of involvement is essential, which means participation and integration of disciplines, levels of government, natural resources management agencies and civil society groups. The world works in an integrated fashion – so

must planning for IWM. Local watershed management plans are often built from the ground upward. But in the process, it is important that the local leaders and other stakeholders understand the nature of a watershed and how it can be managed. Therefore, local community organizations bring people together for educational workshops and other forms of interactive transferring of knowledge (Box 4.2). Ultimately, effective planning and management is dependent on local cooperation and participation.

- Long-term and committed support and investment from key public officials and civil society groups are essential. Good leadership – which includes empowerment of others and application of the subsidiarity principle – is a key as are the availability of resources sufficient to create the capacity to develop, manage and implement often complex plans. The emphasis here is on the long term since successful results in the management of watersheds and river basins is not a short-term undertaking. IWM must be made a long-term political priority.
- Taking advantage of timing and opportunities as they arise is essential. In other words, flexibility is a key to success as circumstances, stakeholders and other factors change over time.
- Plans without their implementation are meaningless. Too often, plans with regard to what needs to be done are formulated and even agreed upon among different stakeholders, but then they are not implemented because an implementation plan

---

Box 4.2. Local leaders invited to watershed workshop. (From *Charlotte Sun Herald*, 12 October 2006. Available at: http://www.sun-herald.com/NewsArchive4/101206/tp3de3.htmldate = 101206& story = tp3de3.htm)

Summary: *A Peace River Watershed Workshop will be held on 20 October 2006, in DeSoto County to provide an overview of the watershed and info on what residents can do to protect the environment and enhance their quality of life.*

ARCADIA – Local community leaders curious to know more about the Peace River Watershed and how they can protect it should circle the date of 20 October on their calendars.

A free workshop will be held at the DeSoto County extension office at the Turner center, 2150 N.E. Roan St., in Arcadia that day.

The Peace River Environmental Education Network and the Watershed Resource Center will host this informative workshop, according to a WRC press release.

Residents can learn what they can do to protect the environment, enhance their quality of life and get a basic overview of how the watershed works, according to the WRC. Questions such as how residents impact natural resources also will be answered.

Presenters from various local and regional organizations, including the Southwest Florida Water Management District and the DeSoto County and Hardee County health departments, will teach about available resources to help with neighbourhood projects. The Peace River Basin Board of the Southwest Florida Water Management District, commonly known as Swiftmud, is supporting the workshop.

The workshop is geared toward community leaders who are involved with homeowner associations, social groups and environmental professionals and organizations, the WRC states.

---

does not exist. Modern society is often too enamoured with the beauty of plans and ignores the hard work of implementation. The how is fully as important as the what and why in planning. Therefore, time and effort need to be spent on designing the best implementation scheme and the best way to gain public support and participation is implementation.

- Treating water as an economic good and planning accordingly is crucial. The consideration and use of water pricing methods, valuation techniques, payment for environmental services approaches and selective privatization of services can be important steps towards making a workable plan.
- Information, education and research are essential inputs. No matter how smoothly the policy issues are resolved, the planning result can be a disaster if poor technical information is used.
- Start small, build on successes, and work upward to a larger scale but work at different levels simultaneously as experience gained is critical. Once again, this relates to flexibility and the ability to see opportunities as they arise. We start small in most cases because of uncertainty of outcomes. We can learn from a small failure or setback, while a major one can end the spirit that enabled the IWM effort to move ahead.
- Environmental, economic and social values can be compatible. The key is to identify the win-win situations and then to work out the trade-offs where conflicts exist. Using appropriate trade-off mechanisms – particularly market incentives – it is possible to arrive at satisfactory solutions to most land- and water-use conflicts. The assumption here is that tenure issues have been resolved. If this is not the case, then this resolution comes first because market incentives do not work well in situations of uncertainty in property rights.

Most of the studies drawing lessons from experience reached the same conclusions but expressed them in different ways depending on the perspective of the institutions involved. The lessons have often been learned the hard way – by failure partly because the experience initially was not shared among practitioners and partly because there was no systematic effort to gather the lessons from the many ongoing watershed management experiences. The question here is, 'What are the practical implications for moving ahead with planning of IWM programmes?'

Planning Elements

Deriving from past experience and the lessons discussed above, one can identify a set of elements in planning for IWM that is practical and workable. Many of these elements involve continuous attention and change, and, therefore, there will be no ending date to the activities involved – for example, in terms of identifying, contacting and informing stakeholders. Furthermore, these elements will constantly be changing as new businesses and enterprises come onto the watershed, new officials are elected and populations of people on the watershed change, so the process of contacting and informing them will also have to be continuous.

The first step is to derive a consensus vision of what the stakeholders need and want from the watershed landscape. This vision is established through appropriate channels, where the process may be steered by a government agency or panel, an ad hoc citizen advisory or planning group, a river basin commission or association and/or other institutional mechanisms. Execution of this step depends partly on the nature of the government involved and partly on the size of the watershed or river basin for which the plan is being developed. Whatever mechanism is used, another of our lessons in planning is that widespread participation, transparency and clarity are needed to ensure that a real consensus and not only a top downward vision is presented to the planning team. State of the art technology transfers need to be used to make sure that information flows effectively and in a timely fashion.

Jones (2001) and others suggest that in addition to the consensus vision there is need for establishing institutional mechanisms and creating the appropriate organizational arrangements *before* actual planning starts. In many cases, however, it might be necessary to move ahead with the planning activities with only temporary institutional mechanisms, particularly if there is no widespread agreement on what mechanisms are needed.

An argument can be made for waiting and building the organizational arrangements for implementation around the plan that emerges – if that is possible. In any case, since implementation and participation in its design will generally involve existing agencies dealing with water, transportation, energy, agriculture, industry, health and so forth, it can be difficult to know where further organizational change is needed and/or how much change is necessary until the planning has progressed and consensus is in sight. The answer will depend largely on the ability of existing agencies to adapt and adjust to new policies and needs and that (in turn) will depend on the policy changes that take place and the policy instruments that are or will be used to encourage people's participation in the planning and implementation processes envisioned in the plan. Three types of instruments are possible:

1. Regulatory mechanisms – restrictions on land and water uses, quotas and controls on management of riparian vegetation and so forth;
2. Fiscal and financial mechanisms – taxes, subsidies, payment of environmental services and so forth;
3. Public investment and management and control of key facilities and resources.

With the vision of what the stakeholders need and want from the watershed landscape clearly in mind, the planning process then involves the following sequential steps (Brooks *et al.*, 2003):

- Monitor and evaluate past activity and identify problems and opportunities in terms of moving towards the vision.
- Identify main characteristics of the problems and/or opportunities and define constraints to overcoming problems or taking advantage of opportunities – develop strategies for action.
- Identify alternative actions, organizational mechanisms and policy instruments to implement strategies given the constraints confronted.
- Appraise and evaluate the design and impacts of alternatives including environmental, social and economic effects – assess uncertainty associated with results.
- Rank or otherwise prioritize alternatives and recommend action when recommendations are requested.

Planning deals largely with the following questions. How do we decide that something needs to be done? What is it we want and need to do? What alternatives do we have available to do it? Which alternative(s) is (are) best, given the circumstances? How shall we go about choosing the alternatives and implementing them?

Parallel policy design

It is likely that a parallel policy design process is ongoing, that is, a process of identifying and then assessing associated policy issues and designing alternative means for their resolution. As mentioned, IWM planning must take place within a policy context, and that context most likely will be changing for various reasons as the planning proceeds (see Fig. 4.1). Also parallel to the planning process, there most likely will be an institutional assessment and government proposals forthcoming on suggested organizational changes. This process will often be helped along by the urging of various civil society groups – often lobbying type groups that are pushing one or another of the stakeholder interests. The types of group that might be involved were briefly discussed in Chapter 3. With all these parallel processes ongoing, the context of the planning effort can become confusing. This has been the experience in the past in IWM planning activities. Therefore, the synthesis of lessons put forth earlier in this chapter is important – particularly the lessons concerning long-term commitment, clarity of objectives and process and good communication.

One thing we know is that the elements included in a given planning exercise will depend largely on what type of watershed or river basin we are dealing with – how important it is in terms of the local, regional and national economy – and what the politics are in terms of those who are promoting IWM planning. Many small local communities establish watershed committees or advisory groups and initiate an informal process that might or might not grow into something major. On the other hand, as in the case of the Sacramento-San Joaquin Delta, the Governor of the State of California initiated the formal process at the state level (Box 4.3) with the establishment of a blue ribbon task force to establish the vision and strategic plan.

Planning Tools

Below the surface of the seemingly straightforward planning process described above, there lies a more complex set of specific actions and activities that need to be carried out to achieve useful results. As someone once said, 'the devil is in the details'. These actions and activities include:

1. Prioritization of issues, problems and opportunities;
2. The design of options to address problems and take advantage of opportunities;
3. The assessment and appraisal of options and activities;
4. The design and assessment of implementation options;
5. Formulation of recommendations and decisions for choice of options and implementation strategies.

Fortunately, these activities are involved in most planning processes and not only those in IWM planning. As a consequence, a number of useful tools have already been developed to deal with them. The key with IWM – and what sets it apart – is the conscious attempt to be holistic, recognizing that most activities on a watershed affect or depend on other activities. IWM planning is multisectoral, cross-disciplinary and an integrative and a dynamic process.

Box 4.3. Sacramento-San Joaquin Delta Estuary Planning.

Recognizing the importance of the Sacramento-San Joaquin Delta Estuary to the State of California, the Governor of California, on 28 September 2006 issued Executive Order S-17-06, the details of which state:

1. *I hereby initiate the Delta Vision and establish an independent Blue Ribbon Task Force to develop a durable vision for sustainable management of the Delta. Making the Delta more sustainable will require a concerted, coordinated and creative response from leaders at all levels of government, stakeholders, academia and affected communities, and will require significant private and public partnerships and investments. The Delta Vision is designed to accomplish these goals:*

(a) *Meet the requirements of Assembly Bill 1200 (Water Code Sections 139.2 and 139.4), Assembly Bill 1803 (Water Code Section 79473) and SB 1574.*

(b) *Coordinate and build on the many ongoing but separate Delta planning efforts.*

(c) *Assess the risks and consequences to the Delta's many uses and resources in light of changing climatic, hydrologic, environmental, seismic and land-use conditions. This assessment will look at:*

- *The environment, including aquatic and terrestrial functions and biodiversity.*
- *Land use and land-use patterns, including agriculture, urbanization and housing.*
- *Transportation, including streets, roads, highways, waterways and ship channels.*
- *Utilities, including aqueducts, pipelines and gas/electric transmission corridors.*
- *Water supply and quality, municipal/industrial discharges and urban and agricultural runoff.*
- *Recreation and tourism, including boating, fishing and hunting.*
- *Flood risk management, including levee maintenance.*
- *Emergency response.*
- *Local and state economies.*

(d) *Develop a programme for sustainable management of the Delta's multiple uses, resources and ecosystem. Sustainable management of the Delta means managing the Delta over the long term to restore and maintain identified functions and values that are determined to be important to the environmental quality of the Delta and the economic and social well being of the people of the state. As part of the Delta Vision, priority functions and values will be identified, and measures necessary to provide long-term protection and management will be evaluated.*

(e) *Develop a strategic plan to implement findings and recommendations for public policy changes, public and private investment strategies, Delta-Suisun preparedness and emergency response plans for near-term catastrophic events, levee maintenance options and how to monitor and report performance.*

(f) *Develop recommendations on institutional changes and funding mechanisms necessary for sustainable management of the Delta. Recommendations may include a discussion of oversight, land use and implementation authorities.*

(g) *Inform and be informed by current and future Delta planning decisions such as those pertaining to the CALFED Bay-Delta Program, Bay Delta Conservation Plan, Suisun Marsh Plan, Water Plan, updates of related General Plans,*

Continued

Box 4.3. *Continued*

*transportation and utilities infrastructure plans, integrated regional water man-
agement plans and other resource plans.*

2. *The Secretary of the Resources Agency as chair, and the Secretaries of the
Business, Transportation and Housing Agency, Department of Food and Agriculture
and the California Environmental Protection Agency, along with the President of the
Public Utilities Commission shall be the Delta Vision Committee, for the Delta Vision.
They shall undertake the following:*

> *(a) Explore entering into agreements with private and NGOs to receive funding
> for Delta Vision. In addition, the Director of Finance may also accept monetary
> and in kind contributions to support the activities of the Delta Vision.*
> *(b) Create a Stakeholder Coordination Group to involve local government, stake-
> holders, scientists, engineers and members of the public in this effort to develop a
> Delta Vision.*
> *(c) Select Delta Science Advisors from diverse scientific disciplines to provide
> independent review and advice to the Blue Ribbon Task Force on technical, sci-
> entific and engineering data, analyses and reports.*
> *(d) Report to the Governor and the Legislature by December 31, 2008 with rec-
> ommendations for implementing the Delta Vision and Strategic Plan.*

3. *I will appoint the members of a Blue Ribbon Task Force to include diverse expert-
ise and perspectives, policy and resource experts, strategic problem solvers and
individuals having successfully resolved multi-interest conflicts. The Task Force will
seek input from a broad array of public officials, stakeholders, scientists and engineers. The Task Force will prepare an independent public report that will be submitted to the Delta Vision Committee and Governor that sets forth its findings and
recommendations on the sustainable management of the Delta by 1 January 2008
and a strategic plan to implement the Delta Vision by 31 October 2008.*

Much of this book deals with the nature of the biophysical side of the planning
activities such as:

- How we identify potential technical problems and estimate the biophysical effects
  and environmental impacts of alternative managerial actions on the watershed;
- Assessing runoff, sediment loads, water flows, cumulative effects and the likelihood
  of floods and potential flood damages;
- How we estimate the effects of vegetation removal or additions of vegetation on
  water flows;
- How we assess water quality impacts of varying activities and options.

That a large amount of this book is devoted to the biophysical side of IWM reflects
the fact that technical considerations are mostly specific to hydrology and watershed
management issues, options and activities, while many of the social, economic,
institutional and political issues and concerns are largely common across many fields and
sectors of interest because of linkages between the sectors and fields.

While there are unique data aspects – again, more related to the biophysical production functions and the values assigned to physical inputs and outputs – the tools for the most part are comparatively standard and apply widely such as the procedures for economic analysis that have a prominent role in assessing market incentive mechanisms and values that are appropriate for payments for environmental services and valuation of damages and gains due to watershed management activities. Therefore, in this section, we will consider the socio-economic side of the needed activities including the variety of tools that exist to define, estimate and assess the socio-economic implications of problems and opportunities and the socio-economic impacts and trade-offs of alternative actions. We present an overview for a manager or decision maker who has to deal with the specialists and needs to understand the logic of the processes and strategies employed in doing the analysis to be able to manage the process and evaluate the results. The reader is referred to more detailed materials where specific tools are treated in greater detail for those who might actually use them because the anticipated reader of this book might not be carrying out socio-economic assessments and evaluations and, therefore, does not have much need for the details of such activities.

Economics as a Planning and Management Tool

As much as some people like to think that economic values do not drive most human activity, they do influence to a great extent what people do – and this includes what they do on their land and with the water resources that lie under or flow across their land to the extent that the law permits. Therefore, economics and economic reasoning and assessment are important planning and management tools. Economics help people to assign values to different actions, different outcomes and different impacts and in the process provide a common metric that can be used to compare different alternatives and consider the value of different trade-offs between what may be quite different outcomes. Detailed information on the guidelines for economic appraisals of watershed management projects is found in Gregersen *et al.* (1987); Brooks *et al.* (2003); EFTEC and Environmental Futures Ltd. (2006a,b).

Principles related to economics as a planning and management tool

All socio-economic assessment, appraisal and analysis tools that are relevant for IWM planning depend ultimately on the input of biophysical data, and, as a consequence, the analyses will only be as good as the data that are used to describe the relevant relationships and interactions on the watershed or river basin. False assumptions and poor data can sometimes be dressed up to look good. But such data can also lead to disaster such as when flood plains are not correctly identified and devastating flooding occurs. We cannot overstate the importance of getting the biophysical data and relationships right. No amount of sophisticated analysis and planning can overcome the problems of poor data. With that point clearly made, several caveats or principles need to be kept in mind when one is moving forward with a socio-economic analysis or assessment.

Principle: consider the political and decision maker environment. This principle is fundamental to an economic assessment made in the real world where politics and special interests can dominate decisions and consequent actions. It does no good to spend time and resources on planning for something that does not have needed political support. However, information from assessments can help shape that support in the future and that is a primary reason for doing thorough data gathering and analysis. But it has to be done in a realistic fashion that recognizes the realities of the political environment. Two major types of factors need to be considered in designing an economic assessment (Fig. 4.3). One set is the decision maker considerations which include the objectives for the assessment and the criteria that will likely be used for accepting or rejecting the assessment and its results. The other set of factors are those associated with technical design of the assessment – that is, data requirements in relation to availability, the cost factors, time considerations, reliability issues, reproducibility and acceptance of methods used by professional peer group.

Principle: get the basic evaluation questions right. If we do not get the appropriate questions on the table, then the evaluation effort will either be useless or end up being misleading. There are basic financial and economic questions that are relevant to most appraisals of projects or components of an IWM programme. Financial questions need to be asked because scarce financial resources almost certainly will be involved, while economic questions need to be asked because most watershed man-

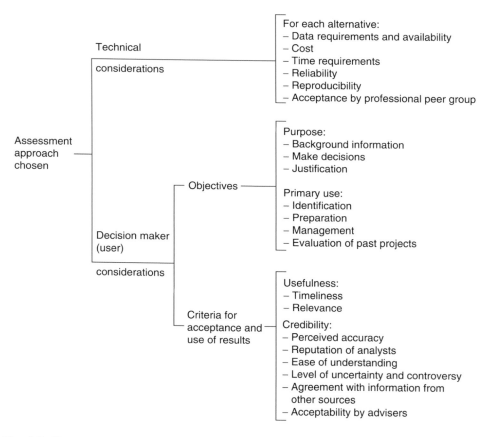

Fig. 4.3. Technical and decision maker considerations in designing assessments. (From Gregersen and Contreras, 1992.)

agement projects, programmes and activities involve public interest and go beyond the private sector in terms of benefits and costs also.

The first question to ask is whether the project is financially acceptable to the relevant interested parties, that is, is the project within the available budget? A second question asked is, over time are the financial returns likely to be sufficient to provide incentive for investment? Still another question is how the financial costs and returns are distributed among the different parties involved and whether this distribution is equitable and acceptable to all the stakeholders in the project to make it successful. If they are not, some form of redistribution needs to be made to make it acceptable. If upstream land users are going to be paid for environmental services, then we need to ask if the amounts proposed are going to be large enough to elicit the wanted actions. This question also relates directly to the equity issue – that is, are all people, including the poor, who tend to be quite helpless, getting their fair share of the benefits and paying the costs that they are deemed capable of paying?

The fundamental socio-economic question – which is the economic efficiency question – that needs to be asked is if the benefits to the nation or to the relevant subunits of the nation affected by the watershed exceed all the costs involved when both are valued in economic terms. We further need to ask if all the components of the project have benefits that exceed costs. If the main objective of the watershed intervention is to increase the aggregate time adjusted economic benefits – goods and services – derived from the use of a country, region or relevant population's limited resources, then this economic efficiency question is the fundamental one that is of concern.

It should be noted that a key to success is that these questions should be clarified early in the planning process and specified in detail. If the responsible analyst merely estimates the financial worth of the project, then specificity with regard to financial worth from whose point of view should be requested. However, if the charge is calculating the economic worth of the proposed activity or programme, then again, more specificity is needed. Costs and benefits from whose point of view must also be known. There could be instances when the project is economically profitable from the point of view of a state that receives resources from the country but not economically viable from the country's point of view. This fundamental principle related to getting the questions correct is related to the next principle.

Principle: analyse separable project components separately. While this principle may seem obvious to most people, it is often ignored. The basic idea is that if separable components and their costs and benefits can be identified, they should be analysed separately since each separable project component should have benefits at least equal to its costs – unless there is a political reason for including components where economic benefits are less than costs. But that fact should then be made clear so that the consequences can be considered. The danger of not separating components is that the net costs associated with one dominant component can lower the benefit of the whole project to a point where it is rejected despite the fact that the rest of the separable project components had higher socio-economic rates of return.

Principle: make a clear distinction between financial and economic efficiency analyses. Financial values are based on actual market prices paid by buyers and received by sellers; therefore, only market traded goods and services are included in the analysis. These values are generally clearly in evidence since they are represented by actual transactions in money. On the other hand, economic values are the total values involved – that is, the real costs of producing or obtaining something and the real total benefit of what is produced or obtained including all of the external effects of a given production activity. This distinction is important when looking at water resources and watershed management since there are many non-market values that come into play. An upstream farmer might divert water into an irrigation project. There are market costs and returns involved for the farmer – perhaps increased revenues to the farmer that exceed his or her costs of diverting the water and creating the irrigation infrastructure. However, other costs are also involved. In general, diversion involves loss of water that might otherwise have benefited downstream water users, and even the water that does return to the river from which it was taken might be more heavily polluted than if diversion had not occurred, which adds a cost downstream.

In the words of economists, the project illustrated involves negative externalities – that is, the costs that are not borne by the farmer and not accounted for in the financial analysis carried out by the farmer. From a financial perspective of the upstream farmer, this could be a profitable undertaking. However, from the society's economic perspective, it might be a loss overall. Similarly, a farmer who practices soil conservation and good riparian management upstream might create positive externalities downstream that never show up in the farmer's financial calculations. In this case, the economic value to society of the soil conservation activity exceeds the financial benefit to the upstream farmer.

There are two significant and basic differences between economic and financial efficiency measures:

1. The benefits and costs – or positive and negative impacts – included in the assessment. Only direct market-traded returns and costs are considered in the financial analysis, while many of the non-market benefits and costs – again, positive and negative impacts – are also included in the socio-economic analysis using economic values.

2. How those costs and benefits (impacts) are valued – A financial analysis always uses market prices, while the best estimates of people's willingness to pay for goods and services are used in an economic-efficiency analysis. Market prices are often adjusted to reflect more accurately the social or economic values in economic analyses. These latter prices are referred to as *accounting* or *shadow prices*.

A financial analysis is not only of interest to individuals and companies in the private sector. The public sector also needs to consider financial aspects of projects in examining the distributional impacts of proposed watershed management project activities – i.e. who actually pays and who gains from such activities. Some of the financial analysis questions that commonly arise in the public sector include:

- What is the budget impact likely to be for the management agencies involved?
- Will the project increase economic and financial stability in the affected regions? Will it have impacts on balance-of-payments?
- Will the project be attractive to the various private entities – such as those upstream? Who will have to put resources into the project to make it work? What will be the income redistribution impacts of the project?

Principles related to values and valuation

The key part of an economic analysis relates to values and valuation. When only market prices are involved as in a financial analysis, it generally is a more straightforward exercise. However, IWM also deals with non-market values that are associated with environmental services representing outputs that are normally not valued in the marketplace. A number of techniques have been developed for estimating non-market values and the true economic as distinct from financial values of goods and services that are traded (cf. EFTEC, 2006a,b; Brooks *et al.*, 2003). The following principles need to be considered when dealing with values and valuation as part of the planning process for IWM.

Principle: use the with and without test in deriving benefits and costs. This is a fundamental principle. The last part of it often is forgotten. It is not enough to say 'before the project erosion was doing $X of damage per year and after the project it dropped to $Y of damage per year'. The with and without principle cannot be forgotten when one is dealing with the dynamic situation that characterizes most of nature. It might well have been that the erosion rate was increasing each year due to cumulative effects and natural processes that built on each other. Therefore, without the project, given the rate of increase before the project, the damage at the time the project started might have been $X but without the project it might have risen to $2X by the time the project benefits of $Y were measured. Only using the before and after figures, net benefits (damage reduction) would have been $X – $Y; but when the with and without numbers are used, the reduction in damage would have been $2X – $Y. This principle relates directly to the next two principles.

Principle: losses avoided are as important as increases in production. In IWM – particularly where many of the main activities are preventative in nature and many of the outputs are losses avoided – it is important to keep in mind that losses avoided are as important as gains in output from an economic point of view. Therefore, the value of 100,000 t of wheat loss avoided due to watershed management interventions that reduce soil erosion and project irrigation water loss is as important as the value of 100,000 t of wheat production gain due to germplasm improvement or other technology improvement. Some planners with the perspective of poor people in mind might argue the details and suggest that the loss avoided might be more valuable because of the price effect of losses versus gains and also because losses can mean disaster for the poor, while equal gains might only mean extra income to spend beyond survival.

Principle: value is created by both supply and demand. Legitimate economic values depend both on supply and demand. It is like a pair of scissors – it takes the two blades to make a cut. Therefore, spending money to save water is a cost, but if that water is not needed and there is no demand for it – for example, if the saving takes place in an environment where there is plenty of water for everyone already – the economic benefit from the water supply increasing activity will be zero. If there is no demand for the additional water, then the additional supply does not matter. When this principle is forgotten, the results can lead to faulty decisions. There could be a demand for the increased water in the future such as through development of hydropower or greater downstream irrigation, but in this case, it becomes necessary to adjust the values involved to account for the difference in timing of supply and increased demand – this point relates to the next two principles.

Principle: the long-term matters and is a critical variable. Sustainability has real meaning and is not just a buzzword. Many of the processes that take place on a watershed involve longer-term effects. For example, building on a flood plain might not lead to disaster for many decades or it could happen next year. Drawdown of the groundwater table could take 20 years with no noticeable effect other than that the wells have to be dug deeper. Then, one year the water is gone. Therefore, uncertainty and risk – the need to forecast and work with models on the biophysical side of the planning process – is essential.

Principle: a dollar of cost or benefit that occurs at one point in time does not have the same value as a dollar that occurs at some other point in time. The logic of this principle is intuitively clear: if I asked you to give me $100 today and I would give them back to you in 10 years, you would ask how much interest are you going to pay me? Everyone tends to have a preference for having their money in hand now – unless someone is willing to pay interest. Poorer people normally have a much greater preference for today since they need to spend their money right now to survive. Economists and financial people have developed an elaborate system of equations to let them equate the value of a dollar in the future back to the present or to any other date closer to now. The basic process is called *discounting*. Bringing values forward in time is also done and that is called *compounding*.

Some of the basic reasoning and fomulae for financial and economic analyses are presented in Annex 4.2.

Principles related to externalities, transfer payments and payments for environmental services

In economics, the term *externalities* is used to denote costs or benefits borne by third parties outside the purview of the activity or project creating the external costs or benefits. They are not borne by those involved in the project or activity; indeed in many cases the latter do not even know about them. A negative external cost is created when, for example, some activity along a river dumps pollutants into the river and they eventually harm someone downstream. Similarly, a positive externality occurs when someone produces some positive benefit that is gained by some non-project entities. For example, a farmer reduces loss of soil on his farm and the reduction in erosion and sediment transfer downstream helps some land or water user downstream. Economists generally advocate attempting to internalize both the negative and the positive externalities; and several resulting principles become very relevant in the economic analysis of IWM projects.

Principle: try to design projects so that 'polluter pays' and 'resource-user pays'. These two principles relate directly to attempts to bring IWM more in line with a market model of behaviour where people pay for what they get; and if they harm others – e.g. through pollution, they pay others for the losses. Some of the latter costs such as taxes on production or deposits are internalized in the cost and, ultimately, the price of goods and services paid by the consumer. This means that these costs need to be estimated and considered in the context of polluters and resource users' willingness to pay. Put another way, these principles focus on the attempt to internalize the externalities associated with watershed management. Such externalities occur because of the strong linkages that exist, such as between upstream and downstream land, water and other natural resources uses, as well as between current and future generations and their abilities to use this land, water and other natural resources. Property rights and how they are treated by government authorities fit prominently into the equation related to externalities.

Principle: payments for environmental services are not a subsidy but a legitimate tool for dealing with non-market values. Related to the principle above, when dealing with the non-market values mentioned above – those that are taken into account in the economic as distinct from the financial analysis – increasingly, governments introduce quasi-market instruments, such as payment for environmental services (PES), to provide incentive for private landowners to help internalize the positive externalities, companies and communities to practice sound IWM and avoid socially damaging actions. The distinction between the payment for environmental services and the older subsidies or handouts given to landowners and users to provide incentive for socially desirable actions is significant in principle. PES provides recognition that landowners are providing – as in a market economy context – legitimate services to specific groups in society that need to pay for these services. People's interest in PES is growing rapidly (Pagiola and Platais, 2002). One point stressed is that appropriate institutional mechanisms need to be established to make such quasi-markets work (Box 4.4). Many examples of actual payment for environmental services schemes are referenced on www.flowsonline.net (see also Tognetti *et al.*, 2004).

Steps in the economic assessment process

Once the basic objective and principles illustrated above are clearly in mind and if the appropriate data are available or can be generated without too much effort, then the economic assessment process itself is straightforward and simple. There are four basic steps in this process:

1. Define and quantify the physical inputs and outputs involved and create tables that show inputs and outputs as they occur over time.

Box 4.4. The process of designing a system of payments for environmental services. (From: World Bank. n.d.)

Once the environmental services have been identified, along with the quantity of them produced and their value, then the process of designing a system of payments for environmental services involves several steps:

1. Charging service users – How can payment systems be financed?
2. Paying service providers – How are payments actually to be made in order to achieve the desired change inland use sustainability efficiently?
3. Creating an appropriate institutional framework – What are the institutional preconditions for the payments to be possible?
The answers to these questions will of course be largely country-specific in their details. Indeed, the nature, extent and value of environmental services are likely to be not just country-specific, but site-specific within a country. Flood protection benefits, for example, depend on what is being protected in any given watershed and on the size and characteristics of the watershed. Within this diversity, however, there is probably a great deal of commonality across countries in the same region – especially if one could develop a typology of watersheds.

---

2. Determine unit values – including both actual financial–market prices and economic values – for inputs and outputs and estimate likely changes in such values over time, e.g. growth in wages.

3. Compare costs and benefits by calculating relevant measures of project worth and other indices and measures needed to answer relevant questions raised by decision makers.

4. Consider the implications of risk and uncertainty, for example, through a sensitivity analysis, which indicates how measures of project worth might change with changes in assumptions concerning input and/or output values.

The overall process is illustrated in Fig. 4.4 which is based on the work of Gregersen and Contreras (1992) that goes through the process in detail. Examples of the application of the process are provided by Gregersen *et al.* (1987) and Shuhuai *et al.* (2001). Annex 4.3 provides a listing of selected readings on valuation and economic analysis of natural resources and integrated water and watershed management for those readers who want to pursue the actual process in greater detail. It also provides a list of useful web sites.

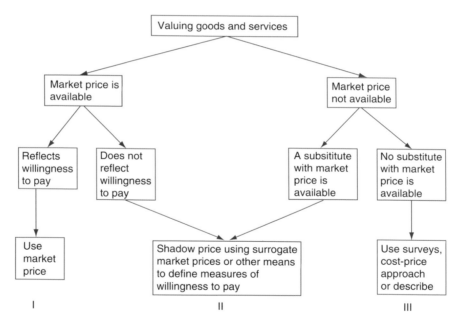

Fig. 4.4. Valuation conditions and approaches. (Gregersen *et al.*, 1987.)

5 Hydrologic Processes and Technical Aspects

Introduction

Surface water flows from landscapes into stream channels within watershed boundaries. The use and management of land and natural resources within watershed boundaries affect the quantity and timing of water flow and the water quality of streams that leave each watershed. Streams from multiple watersheds flow into rivers, with flow at the mouth of the river from the total area of all contributing watersheds being a river basin. The areas that contribute to groundwater, which underlie watersheds, do not necessarily share the same watershed boundaries, although land use on watersheds can influence the recharge and quality of groundwater. People likewise can deplete or otherwise alter groundwater connections to streams, rivers and lakes through pumping of wells and by altering surface water–groundwater interfaces, e.g. on wetlands and along stream channels. Not only does this chapter focus on watersheds and surface water, but also discusses relevant surface water–groundwater processes and issues that affect, and are affected by, people on a watershed.

Watershed Hydrology

Watershed hydrology is an interdisciplinary science that focuses on the union of land-use activities including forestry, grazing, agroforestry, agricultural cropping, mining and urbanization with the science of hydrology. Understanding the hydrologic response to land use and management is paramount if we are to sustain land and water resources for future generations. Gaining this understanding requires that we are able to recognize how human use of land and natural resources alters fundamental hydrologic processes and, furthermore, how these changes in processes translate into water flow response from watersheds. A question one might ask at this point is 'who' should have this understanding. We suggest that these are the people using land and water and those who are responsible for planning and managing these resources.

People who use the watersheds to produce food, fodder, fibre and other natural resource products and obtain amenities do not always recognize the implications of their resource use on water and, conversely, how their use of water influences the land-use practices implemented to produce these other resources. Watershed managers, on the other hand, must often address specific questions related to land-use practices. These questions include:

- What land-use activities can take place on watersheds without causing undesirable hydrologic effects such as flash flooding, soil erosion and sedimentation and water quality degradation?

- What land-use alternatives might be implemented to change hydrologic processes or conditions for a beneficial purpose such as increasing water yields, improving water quality and reducing flood damages?

To answer questions such as these, relationships between watershed management practices and the resulting responses in hydrologic processes are analysed by studying the *hydrologic cycle*. The hydrologic processes of the biosphere and the effects of vegetation and soil on these processes must be recognized within the framework of the hydrologic cycle to achieve effective and sustained use of multiple resources on watersheds. This knowledge is essential to natural resource managers for comprehensive planning, implementation and evaluation that are needed to achieve long-term natural resource and agricultural development.

Hydrologic Cycle

The pathways by which water moves on earth and the processes that affect these pathways are depicted by the hydrologic cycle (Fig. 5.1). The hydrologic cycle is driven by solar radiation, which is the source of energy for evaporation from water bodies and landscapes. The energy consumed in the evaporation process is released when atmospheric water condenses and falls to the earth as precipitation. Precipitation reaching the earth's surface at elevations above sea level then begins a journey back to sea level. Vegetative cover, geology, soils and topography all influence the pathways and timing by which water moves in this often-tortuous journey.

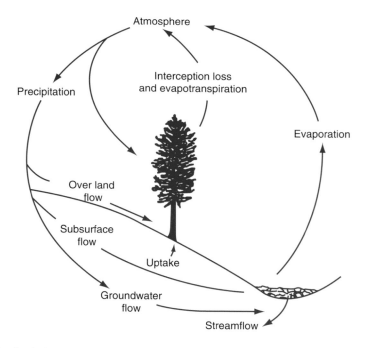

Fig. 5.1. The hydrologic cycle.

Vegetation can catch a portion of falling rain or snow, part of which can evaporate back to the atmosphere, a process called *interception*. This is where a vegetative cover initially affects the fate of precipitation. Water reaching the soil surface can either enter the soil surface, a process of *infiltration*, it can run off or it can pond on the surface of the soil when the rate of infiltration is exceeded. Water that infiltrates into the soil takes different pathways that are determined largely by soil characteristics and the amount of moisture held in the soil at the time of infiltration. Water that infiltrates into the soil can be stored if the soil-water holding capacity, called the *field capacity*, has not been exceeded. Field capacity is the maximum amount of water that the soil can retain under the influence of gravity and is largely determined by the soil texture. Finely textured soils such as clays and clay-loams have numerous small pores that hold large quantities of water in contrast to sandy soils that have larger pores but lower total porosity. Water stored in the soil can be evaporated from the soil surface or taken up by plant roots and, eventually, evaporated from plant leaves into the atmosphere by the process of *transpiration*. The total evaporative loss from watersheds is the sum of interception, evaporation from soil and water bodies, and transpiration, and is collectively called ET.

Once field capacity is exceeded, additional infiltrated water moves downward under the influence of gravity, a process called *percolation*. Percolated water that reaches groundwater moves through the pores in *aquifers*. Aquifers are water-bearing soils or geologic strata that store groundwater. Water moves in aquifers in response to hydraulic gradients, in all cases moving from areas with a higher energy state to a lower energy state. Sometimes, percolated water reaches an impeding layer of soil or rock material that causes water to pond and form *perched groundwater*. In the case of hillslopes, percolating water that reaches large subsurface pores and cavities can flow more quickly than within the soil matrix, eventually emerging from hillslope surfaces as a seep or reaching the toes of hillslopes and discharging directly into a stream channel or other water body. The velocity of subsurface water flow is determined largely by the size of pores in the soil. Pore size is determined by soil texture and structure and by the type of vegetative cover. Most deep forest soils have large pores with high infiltration capacities that promote high rates of subsurface flow. Surface runoff, or *overland flow*, rarely occurs in such soils, but is the dominant pathway by which water leaves the landscape in urban areas, road surfaces and overgrazed pastures.

Overland flow generally occurs at faster rates than subsurface flow through soils – the steeper the slope and smoother the surface, the faster the rate of overland flow. *Subsurface flow* moving through hillslopes (in turn) occurs faster than groundwater flow. Water entering groundwater can take weeks, months or years before reaching major rivers and, ultimately, the ocean. Some of this water can be stored in deep aquifers for millennia. The proportions of excess rainfall or snowmelt that flow as surface, subsurface or groundwater determine the timing of water volumes reaching the stream channel from the watershed landscape. As a consequence, the quantity of water flowing through each pathway determines the timing of streamflow response on the watershed.

Stream channels in arid regions might receive large portions of overland flow from rainfall but little or no groundwater flow throughout the year. Without groundwater contributions, these are *ephemeral* or *intermittent* streams, meaning that they flow only for a short time following rainfall or snowmelt events. Streamflow can occur year-round in areas where groundwater feeds into a stream channel. The stream is *perennial* in such instances. Streams that receive a large amount of surface runoff from compacted soils or impervious urbanized areas will be flashy, with water levels rising quickly in contrast to those that receive predominantly subsurface and groundwater flow. Water flows more slowly where subsurface and groundwater flow dominates as it does on the surface where the topography is flat. Water flows into, through and out of wetlands, lakes and ponds slows the travel time through the watershed to the stream channel. This situation also favours higher losses of water through evaporation and transpiration. The combination of retention and detention storage in a watershed, therefore, influences the quantity and timing of water flow into stream channels. Once in a channel, the water flows in response to the channel gradient (slope) and other stream channel conditions, ultimately, returning to the sea; hence, the cycle is complete.

Water flow in stream channels is seldom constant, with high flows during periods of rainfall or snowmelt and low flows in dry periods. Floods occur when the water level in the stream channel exceeds its adjacent streambanks, a natural process that occurs annually for many streams. Stream channels develop their characteristic form or morphology because of stream gradient, geologic and soil materials through which the channel flows, and the historic streamflow patterns.

Water moves slowly in flat meandering streams and quickly in steep mountainous channels. As the velocity and quantity of flow increase with time, the power of streamflow to pick up and transport sediment (eroded soil particles) also increases. Sediment transport, therefore, occurs naturally in streams and rivers, with higher levels of sediment transported by large flows, particularly during flood events. As a result, streams and rivers alter the watershed landscape by cutting downward into the earth in upper portions of river basins, meandering within valleys and depositing sediment on flood plains and estuaries. Land-use practices that increase the magnitude of water flow for a given amount of precipitation also increase stream power and with greater capacity to transport sediment, can accelerate these processes causing increased channel degradation in upper areas and greater sediment deposition downstream.

Water budget

The hydrologic cycle is viewed as a series of hydrologic components. Watersheds or river basins can be characterized by arranging these components into a *water budget*, which simplifies the complexity of the hydrologic cycle somewhat by separating all water components into inputs, outputs and storage components (Fig. 5.2). The water budget is essentially an application of the *conservation of mass* principle to the hydrologic cycle. It serves as an accounting procedure which quantifies and balances hydrologic components on a watershed basis (Dunne and Leopold, 1978; Gordon *et al.*, 1992; Satterlund and Adams, 1992; Brooks *et al.*, 2003; Chang, 2003). Most land-use activities that influence the hydrologic cycle can be considered in terms of influencing the magnitude of the storage components in a water budget or how the flow components are altered by the activities. To illustrate the concept of a water budget for a particular watershed and a specified time interval:

$$I - O = \Delta S \qquad (5.1)$$

where I = sum of all inputs of water to the watershed (precipitation, groundwater flow from adjacent areas); O = sum of all water loss or outputs from the watershed (streamflow from the watershed, water that seeps into deep groundwater, and all

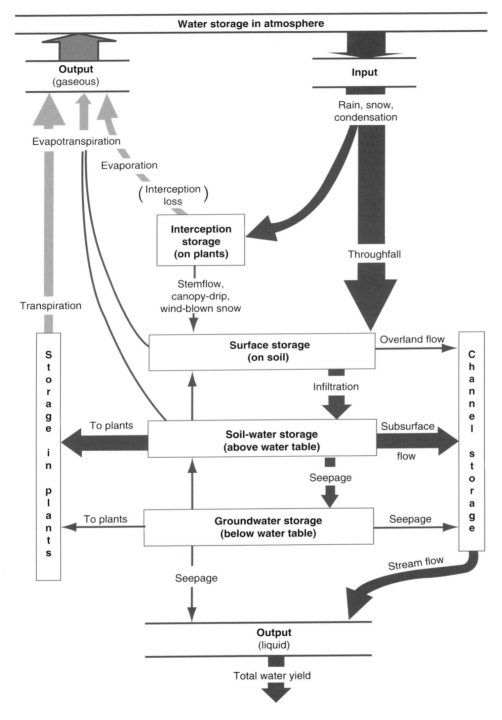

Fig. 5.2. Inputs, outputs and storage components of the hydrologic cycle of a watershed. (Adapted from Anderson *et al.*, 1976.)

forms of evaporation losses); ΔS = change in storage of the volume of water in the watershed, that is, storage at the end of the time period (S_2) minus storage at the beginning of the time period (S_1); all changes in storage in the watershed are included, such as soil moisture, lakes, reservoirs, ponds and wetlands.

Application of the water budget

The water budget is a useful tool that can be used to characterize the overall water status of watersheds or river basins and study their hydrologic behaviour. Substituting with the components of the hydrologic cycle (in units of volume per unit time), Equation (5.1) becomes:

$$P + L_i - ET - L_o = \Delta S \tag{5.2}$$

where P = precipitation; L_i = groundwater that enters the stream channel from an adjacent watershed; Q = streamflow at the watershed or river basin outlet; ET = evapo-transpiration losses; L_o = leakage out of the watershed by deep seepage to groundwater that enters a channel elsewhere; ΔS = change in storage volumes over time ($S_2 - S_1$). Annual water budgets are commonly used to characterize a watershed, river basin or a large land mass such as a continent. Annual water budgets for watersheds or river basins are used for hydrologic analysis because of the simplifying assumption that changes in storage in the watershed system in a year are relatively small. Annual water budget computations are often made beginning and ending with wet months or dry months for a given watershed. The difference in storage between the beginning and end of the year in either case is relatively small and can be assumed to be negligible for most calculations. With this assumption, ET, which cannot be adequately measured in the field, can be estimated from measurements of precipitation and streamflow for a watershed over a year as follows:

$$ET = P - Q \tag{5.3}$$

Provided that an acceptable measurement of precipitation is obtained, a second assumption made in applying an annual water budget is that the total outflow of water from the watershed has been measured as streamflow. This implies that there is no loss from, or gain to, streamflow at the watershed outlet by deep seepage or groundwater inflow associated with the underground geological strata from adjacent watersheds. In some watersheds, particularly in arid regions of the world, portions of streamflow can percolate into the channel bottom or sides, referred to as *transmission losses*. Some of this water can enter groundwater aquifers and leave the watershed as groundwater flow. The surface boundaries of the watershed do not coincide with the boundaries governing the flow of groundwater in these cases. Such occurrences are common when geologic strata such as limestone underlie a watershed. In such instances, there are two unknowns in the water budget, ET and groundwater seepage (L), which result in:

$$ET + L = P - Q \tag{5.4}$$

When it is not appropriate to assume that the change in storage is negligible, it must be estimated in some way. While this is a difficult task, changes in storage are often estimated by periodical measurements of the soil water content on relatively small watersheds, and also by changes in storage in wetlands, ponds, lakes and reservoirs.

Quantifying components in the hydrologic cycle facilitate the use of a water budget in studying the hydrologic behaviour of a watershed before and after it has undergone changes in land use or management. This exercise is helpful in assessing the hydrologic impact of such changes. Methods of measuring the various climatic and hydrologic entities needed to perform water budget and other hydrologic analyses are described in hydrologic references (Lee, 1980; Maidment, 1993; Brooks *et al.*, 2003; Chang, 2003).

Continental water budgets can provide insight into climatic conditions and, when adequate knowledge of the respective hydrologic components is available, indicate the relative abundance or shortage of water for human use (Box 5.1). A major difference between a water budget for a continent and that for a watershed or river basin is largely that of scale. Evaluations and the interpretations of water budgets at a continental scale must be made with this condition in mind.

Application of the water budget as a hydrologic method is illustrated for different land-use changes and climatic regions throughout this chapter. A simplified example of applying a water budget to estimate ET on a watershed basis is presented in Table 5.1. In this example, average annual precipitation and surface runoff for the undisturbed forest were 635 and 75 mm, respectively, leaving 560 mm for ET. Following the partial cutting of the forest, the annual streamflow runoff increase of 25 mm is attributed to a reduction in ET of 25 mm. The annual precipitation input of 635 mm

---

Box 5.1. Water budgets at a continental scale: some considerations.

Although the average annual precipitation (P) for the continent of Africa is similar to that of Europe, the proportion of P that is consumed by ET is much higher for Africa (80% compared to 57%, respectively), indicating the disproportionate area of arid regions in the continent. Water budgets over large areas provide little insight into the reality of water resource status, because it is the distribution of water that is of paramount importance. Annual streamflow in Africa is the lowest of all continents (listed), even though the Congo River – with its basin of more than 4 million square kilometers – produces streamflow that is second only to that of the Amazon River in South America. If we considered individual watersheds across Africa, watersheds in arid regions would have low annual P, high ET, leaving little or no residual water for streamflow (Q), whereas in the humid Congo Basin, streamflow would represent a large portion of annual precipitation.

Examples of continental water budgets (van der Leeden *et al.*, 1990).

	Annual P (mm)	Annual Q (mm and % of P)	Annual ET (mm and % of P)
Europe	734	319 (43%)	415 (57%)
Africa	686	139 (20%)	547 (80%)
Asia	726	293 (40%)	433 (60%)
Australia	736	226 (31%)	510 (69%)
North America	670	287 (43%)	383 (57%)
South America	1648	583 (35%)	1065 (65%)

Table 5.1. Average annual precipitation, ET, and runoff for undisturbed forested and harvested (partial cut) watersheds in the ponderosa pine forests of the south-western USA. (From Ffolliott and Thorud, 1977.)

Water budget component	Undisturbed (mm)	Partial cut (mm)
Precipitation	635	635
Surface runoff	75	100
ET	560	535

remained the same as precipitation input is generally not affected by modifications in the structure and composition of vegetative cover on watersheds. Annual runoff increased by 33%, while the annual ET decreased by 4.5%. In this example, a small reduction in ET can cause a substantial increase in runoff or streamflow.

Impacts of watershed characteristics on hydrologic processes

The climatic regime and the vegetative, geologic, topographic and soil characteristics of a watershed determine how the hydrologic processes of the watershed respond to precipitation inputs. Hydrologic processes on watersheds that undergo land-use activities are influenced principally through changes in the type and extent of vegetative cover, the physical and biological characteristics of soils and changes in watershed features such as wetlands, stream channels, riparian corridors, lakes, ponds and reservoirs. Likewise, the extent and persistence of anthropogenic changes such as urbanization influence the magnitude and persistence of the hydrologic response. Natural variations in climate and precipitation – in addition to the occurrence of extreme events such as droughts and hurricanes – can mask or accentuate the effects of land use. We recognize that watershed hydrology is a *complex science* in this context. However, to gain insight into hydrologic responses to land use, we will discuss the manner by which land-use activities affect key hydrologic processes and, ultimately, streamflow quantity and pattern. Implications for groundwater and water quality are also considered.

Precipitation

The influence that vegetation, and particularly forests, exerts on climate and precipitation has been controversial for many centuries (Andreassian, 2004). Although the claims of macroclimate influence have been largely dispelled, there is little doubt that vegetation can influence the microclimate, particularly through local effects on air humidity, wind and temperature.

Forests occur where there is an abundance of precipitation – a fact that has led some people to think that forests attract precipitation. As noted by Lee (1980), '…the natural coincidence of forest cover and higher precipitation has undoubtedly caused, or at least reinforced, the popular notion that forests increase or 'attract' rain…'.

However, the argument that forest ecosystems exert a great influence on the quantity of precipitation that falls on a watershed is weakened when assessing the mechanisms by which forest vegetation could conceivably have such an effect and when considering the role of ocean evaporation globally. For example, the removal of all of the forest cover in the world would only reduce global precipitation by 1–2% at the most (Lee, 1980). On a smaller scale, Calder (2005) suggests that deforestation of a watershed has little effect on regional precipitation, although exceptions could occur in river basins where rainfall events depend largely on internally driven circulation patterns such as the Amazon Basin. However, even here, complete deforestation and replacement with non-forest vegetation would have minimal effects on the overall basin rainfall (Brooks et al., 2003; Calder, 2005).

In some climatic regimes, forests can enhance moisture inputs to watersheds by intercepting fog or cloud moisture. Although not a precipitation process, interception of atmospheric moisture by *cloud forests* in coastal regions and high mountainous areas – where there is a high incidence of fog or low clouds – can add water to watersheds that would not otherwise occur as precipitation. Intercepted cloud moisture by tropical montane cloud forests can add up to 4 mm/day of water to watersheds. Contributions of fog or cloud moisture vary from 5% to 20% of the normal rainfall, but they can exceed 1000 mm/year in some locations (Bruijnzeel, 2004). Fog drip from the moist temperate Douglas-fir forests on the windward side of the Cascade Mountains in the Pacific Northwest region of the USA can also contribute moisture to watersheds, although such contributions are also variable (Harr, 1983; Ingwersen, 1985). Augmentation of moisture to watersheds through fog or cloud moisture can help to sustain streamflows during the dry summer season. Importantly, the removal of cloud forests in either temperate or tropical regions does not affect annual precipitation, but it can reduce the quantity of atmospheric moisture that is added to a watershed. Conversely, if we convert pastures or agricultural croplands to either forests or agroforestry practices under these conditions, we hypothesize that streamflow volumes can increase.

While the type and extent of vegetative cover exerts little effect on the overall magnitude of precipitation that falls to earth, its fate is greatly affected by vegetation. Of all forms of land use and types of vegetative conditions, forests tend to affect the fate of precipitation more so than other vegetative systems. Forest influences are principally expressed though the processes of interception, infiltration and ET processes that can dramatically affect soil moisture storage on watersheds and the amount of liquid water that is yielded from a watershed as either stream- or groundwater flow.

Interception

Although vegetation does not significantly influence the total precipitation falling on an area of land, called *gross precipitation*, the type and extent of vegetation does determine the amount of precipitation that reaches the mineral soil surface, which is called *net precipitation*. If vegetation is not present, gross precipitation equals net precipitation. As vegetative cover and its biomass increase on a watershed, the interception capacity of the vegetation generally increases and, as a consequence, the portion of gross precipitation that becomes net precipitation diminishes. The *interception storage capacity* of a vegetative community is a function of the total surface leaf area, number of vegetative layers, litter and duff accumulations and other vegetative characteristics. Multilayered forest canopies with herbaceous understories and large accumulations of organic litter on the soil surface have the greatest capacity of all vegetative types to intercept and evaporate precipitation.

Each type and age class of a vegetative community has its own, often unique, interception storage characteristics. The total *interception loss* that occurs during a particular rainstorm is also a function of the rainfall intensity, wind velocity and other storm and evaporation conditions. Interception loss during intense rainstorms of short duration is generally a small percentage of the total storm rainfall for any vegetative community. Even a dense mature forest might only intercept a few mm of rainfall out of a 100 mm thunderstorm event. On an annual basis, however, the cumulative interception loss by forests can amount to a large percentage of annual precipitation, with lesser amounts in declining order for shrubs, grasses and annual crops, respectively.

The magnitude and relative importance of forest interception to an annual water budget differs among forest types, structure, density and climatic conditions as indicated (Helvey and Patric, 1988; Ffolliott and Brooks, 1996; Calder, 2005):

Interception by temperate broad-leaved deciduous forests:

- Eastern USA: Interception averages about 12% of the annual rainfall.
- Eastern Europe: Interception averages about 25% of the annual precipitation, varying between 30% and 40% in the growing season and 20% and 30% in the dormant season.

Interception by temperate conifer forests:

- Southern USA: Interception averages about 15% of the total rainfall.
- Rocky Mountains in USA: Interception varies between 10% and 25% of the annual rainfall depending on forest density.
- Pacific Northwest USA: Interception varies with the size of the rainstorm, ranging from 100% in storms less than 1.5 mm to 10–15% in storms greater than 75 mm.
- Canada: Interception varies from 15% to nearly 40% of the annual precipitation, depending largely on the type of precipitation event – rain or snow – and density of the forest cover.
- England: Interception ranges from 25% to 40% of the annual precipitation.
- India: Interception of coniferous and broad-leafed forest plantations averaged 20–25% and 20–40% of the annual rainfall, respectively.

Interception by humid tropical forests:

- Peninsular Malaysia: Interception averages 20–25% of the total annual rainfall amount.
- Java, Indonesia: Interception averages 14–21% of the annual rainfall.
- West Java, Indonesia: Interception for lowland rainforests amounts to about 20% of the gross rainfall.

Interception losses in arid regions of the world are generally lower than those found in wetter regions because of lower rainfall and the lesser canopy densities usually encountered. While the percentage of annual precipitation that is lost through interception is variable, it can represent significant losses in some instances. For example:

- South-western USA: Interception is between 5% and 10% of the annual rainfall in the conifer woodlands.
- Northern Mexico and the south-western USA: Interception can be 70% of the limited summer rainfall in the largely evergreen oak woodlands.

Snowfall interception is more difficult to quantify than rainfall interception, largely because neither the initial amount of snowfall intercepted nor the water content of the accumulated snow on foliage can be measured accurately. Most studies of snowfall interception indicate that snowpack water equivalents are greater in clearings than in the trees (Brooks *et al.*, 2003). Mechanisms contributing to these differences are not always well understood and relate to factors in addition to the interception of snowfall such as wind blowing snow off foliage. Whether snowfall interception represents a significant loss or not in the water balance of a watershed, forests strongly influence the deposition of the snow on the ground by redistributing the accumulated snowpack which can affect the amount of snowmelt runoff from the site.

In addition to representing a part of the total ET loss from a watershed, interception also plays a role of protecting the soil surface from falling raindrops. Vegetation and the organic litter and duff layers covering a soil surface are critical in dissipating the energy of falling raindrops. This protection minimizes the energy available to dislodge soil particles that can lead to soil erosion. Without this protection, the soil surface can become compacted and sealed which can then reduce the infiltration characteristics of the soil.

Evapotranspiration

ET is the sum of interception loss, soil evaporation and transportation. It is the major loss of water from most watersheds in the world and is a function of the *potential evapotranspiration* (PET), the vegetative characteristics of a watershed and the availability of water. PET is determined largely by the energy that is available for evaporation and represents the maximum ET that can occur from an area under existing climatic conditions. When water is readily available to roots (under moist soil conditions), actual ET can approach or equal PET. As a consequence, sites in arid regions generally have high PET but lower actual ET than humid forests or wetlands. ET can exceed 90% of annual precipitation on some forested and wetland sites.

Plant height, leaf area, structure, physiology and phenology combine to explain species differences in annual ET losses. Forests in humid areas usually exhibit higher ET than herbaceous vegetative cover such as grasses and annual crops, largely because their larger surface areas are more exposed to wind and advected energy (Calder, 2005). Conifers often have higher annual ET than deciduous trees because of higher interception losses. The longer growing season of perennial plants tends to result in higher annual ET than that of annual crops – other factors being equal. Deeper-rooted trees and plants in arid regions can extract soil moisture from deeper horizons and sustain higher levels of ET than shallow-rooted plants. There are many exceptions to these rules which can often be explained by plant physiology.

As ET increases, the amount of water that is yielded from watersheds as streamflow or groundwater recharge decreases. ET is influenced by land management practices that alter the type of vegetation and its extent on a watershed (Dunne and Leopold, 1978; Gordon *et al.*, 1992; Satterlund and Adams, 1992; Brooks *et al.*, 2003; Chang, 2003; Calder, 2005). Changes in vegetative cover that reduce ET over a watershed can result in increases in streamflow and/or groundwater recharge in some cases. Water resource managers have shown interest in the potential of manipulating vegetative cover for such purposes in areas of water scarcity. However, when changes in vegetative cover and other land-use practices increase ET, we generally see reduced water flow. Such changes might be desirable in areas where soils are frequently water logged.

Although the significance of ET is recognized, hydrologists are hindered by their inability to directly measure ET from watersheds. Consequently, there is not a good understanding of how the process of ET varies across a landscape or the feedback mechanisms that control ET in natural environments. Much of what we do understand about ET has originated from controlled experiments or estimated from water budget analyses as already discussed. Of all topics in watershed hydrology, ET remains a topic in dire need of more fundamental research that examines both small-scale and large-scale processes.

Examples from research are presented below to suggest the overall importance of ET in the overall water budget of watersheds (Ffolliott and Brooks, 1996; Zhang et al., 1999; Brandes and Wilcox, 2000; Cleverly et al., 2002; Calder, 2005):

ET in temperate climates:

- Western USA: ET in the spring and fall and adjusted for summer precipitation averages 488, 379, and 226 mm on small plots occupied by aspen forest, spruce forest and grassland vegetation, respectively. These ET values are approximately 90%, 70% and 50% of the annual precipitation falling at these sites, respectively.
- Western USA: ET differences among bare soil, herbaceous vegetation and a mixture of broad-leaved tree covered plots in Utah were 285, 390 and 535 mm, respectively, representing nearly 20%, 30% and 40% of the annual precipitation input to the plots, respectively.
- South-western USA: ET by ponderosa pine forests represent from 78% to 95% of annual precipitation, varying from 463 to 835 mm. ET follows the bimodal seasonal pattern of precipitation in the region and is estimated to be 95% of the annual precipitation.
- England: ET by poplar trees averaged 3.6, 3.8 and 2.7 mm/day in the early, middle and late part of the growing season.
- Wales: ET by conifer forests equals 864 mm or 43% of annual precipitation from February 1974 to October 1976, a period in which precipitation was 5444 mm, total ET was 45% of precipitation of which 36% of total ET is transpiration and 64% interception.

ET in more humid climates:

- India: ET by broad-leaved trees and eucalyptus amounts to 1145–2705 mm, and 5525 mm, respectively
- Thailand: ET by dry evergreen forests averages 950 mm annually, while ET in the evergreen forests is about 1035 mm or nearly 50% of annual rainfall.
- West Malaysia: ET in the rainforests is 1065 mm, while ET in areas where rainforests and rubber plantations are intermixed is about 995 mm.
- West Java: ET in wet tropical forests amounts to 1481 mm or 52% of the annual precipitation.
- Brazil: ET in moist tropical forests of the Amazon River Basin is 1393 mm or 54% of the annual precipitation.

- Bruijnzeel (1990) summarized ET by tropical lowland forests as follows:

Location	Annual P range (mm)	ET/P (%) average	ET/P (%) range
Latin America	1684–3751	57	39–81
Africa	1800–5795	63	20–77
South-east Asia	1727–4037	61	35–90

ET in arid climates:

- South-western USA: ET by salt cedar, a phreatophyte plant introduced to the south-western USA, ranges from 1000 to 2200 mm associated with water table depths of 2.7 and 1.5 m, respectively.
- South-western USA: ET by salt cedar on non-flooded and flooded sites totals 740 and 1220 mm/year, respectively.

Infiltration

Infiltration capacity is the maximum rate of water entry into the soil surface and is determined by the soil texture, structure, depth, by the vegetative cover conditions and by the land use that affects surface soil conditions. Because soils in arid regions have little vegetative cover, these soils naturally become compacted directly by the energy of falling raindrops. This raindrop impact can also remove and wash away the smaller soil particles, leaving larger pebbles and rocks which form an almost impervious desert pavement in many cases. Watersheds supporting naturally occurring (undisturbed) plant communities such as forests and grasslands tend to have higher infiltration capacities than watersheds that have been heavily grazed or cultivated.

Natural (undisturbed) forests also tend to have the highest *rates of infiltration* in comparison to other types of vegetation. This high infiltration is a main contributing factor to the idea that forests have a moderating effect on streamflow (Dunne and Leopold, 1978; Gordon *et al.*, 1992; Satterlund and Adams, 1992; Brooks *et al.*, 2003; Chang, 2003). Forest soils are relatively porous and open at the surface, largely because of the litter and duff cover, the high organic content and the large number of macropores that occur as a result of worm and burrowing animal activities. When forest cover is removed and replaced with agricultural crops or other vegetative cover, the hydrologic properties of soils can be changed dramatically. These changes in soil characteristics have serious implications regarding runoff and erosion.

Temperate forests with deep soils generally have infiltration capacities that are often higher than rainfall intensities and are much greater than those generally observed elsewhere. For example (Hewlett and Troendle, 1975; DeBano, 1981; Ffolliott and Brooks, 1996; DeBano *et al.*, 1998):

- Long-term research at the Coweeta Experimental Forest in the south-eastern USA has shown that forest cover with deep soils promotes high infiltration capacities that are far in excess of rainfall intensities that occur there.
- Infiltration measurements in northern Minnesota indicate that infiltration capacities in sandy loam soils with forest cover commonly exceed 200 mm/h. In contrast, infiltration rates on adjacent logging trails and landing areas can be less

than 10 mm/h. Overland flow, therefore, only occurs on forest soils that have been compacted by vehicle traffic and skidding of logs.

- Forests in colder temperate regions also influence the occurrence, depth and type of frost in the soil, which can then affect infiltration rates. The type of forest cover, soil texture, depth of organic litter and the depth of snow (when snow is present) influence soil frost and, hence, infiltration rates. Removal of forest overstories in the north-central USA results in deeper and more frequent soil frost than that which occurs with forest cover. Furthermore, the occurrence of a *concrete frost* is more prevalent in soils that are non-forested than in soils with forest cover. As this term implies, infiltration rates for concrete frost conditions are minimal.

- Soils supporting some vegetation types develop a characteristic of water repellency which can reduce infiltration capacities. Although the occurrence of these *hydrophobic soils* is common, the causes of this condition are not always well known (DeBano, 1981). Most hydrophobic soils repel water as a result of organic, long-chained hydrocarbon substances coating the soil particles. As a consequence of these substances, water *beads up* on the soil and will not penetrate readily. Hydrophobic soils are found in chaparral (sclerophyllous) shrub communities in the south-western USA. Fire intensifies the hydrophobic condition, and, apparently, volatilizes organic substances and drive the water-repellent layer deeper into the soil.

Infiltration rates of soils in the humid tropics are generally higher than soils in non-forested temperate and many arid regions of the world. These relatively high infiltration rates are usually attributed to the more favourable structure and greater stability of aggregates in these soils. For example (Ffolliott and Brooks, 1996):

- Thailand: Infiltration in excess of 1115 mm/h – with a final infiltration rate of 280 mm/h – has been measured in the hill-evergreen forests; in contrast infiltration rates under dry dipterocarp forests averaged 195 mm/h.
- Philippines: Extensive hillslope cultivation and inappropriate farming practices exhibited lower infiltration rates than the original forest cover. Infiltration rates on sites that are burned annually ranged from 45 to 70 mm/h.

Land-use practices such as agricultural cultivation, intensive livestock grazing and road construction can all reduce infiltration capacities in affected areas in contrast to natural vegetation conditions. Compacted soils can take long periods of time before they can recover to undisturbed conditions. Final infiltration rates in north-western Minnesota in the USA averaged 4, 4.5 and 10 mm/h for fallow croplands, 11-year-old hybrid poplar plantations on agricultural croplands and 22- to 34-year-old hardwood forests, respectively (Shank, 2002).

Flow processes

The influence of vegetation and land use on interception, ET, infiltration rates and soil conditions, collectively, affect the pathways, quantity and timing of water flow within and from a watershed. Soils possessing low infiltration capacities promote overland flow where the rate of rainfall or snowmelt frequently exceeds the infiltration capacity. Soils with high infiltration capacities promote subsurface flow. Soil moisture content and the amount of storage space that is available in a soil at the time of rainfall or snowmelt runoff affects both the infiltration rates and volume of water that leaves the watershed in surface or subsurface pathways. When water moves predominantly as subsurface flow rather than as overland flow, there is a moderating effect on flow entering stream channels. By observing and studying the quantity, pattern and quality of streamflow in channels, we see the integrated and cumulative effects of climate and land use interactions with watershed characteristics.

Streamflow

A study of flow pathways and streamflow response of watersheds or river basins is facilitated by the use of a streamflow *hydrograph*. A hydrograph is a graphical

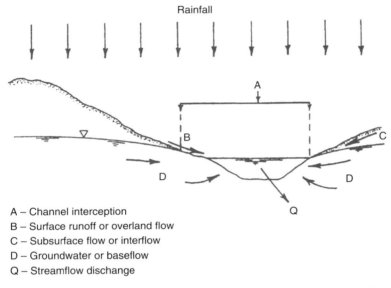

A – Channel interception
B – Surface runoff or overland flow
C – Subsurface flow or interflow
D – Groundwater or baseflow
Q – Streamflow dischange

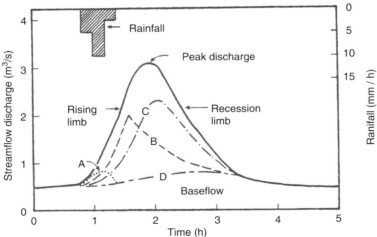

Fig. 5.3. Sources of flow to a stream channel and the corresponding hydrograph. (From Brooks *et al.*, 2003.)

representation of streamflow in volumes per unit time (m³/s) plotted with time at some point in a channel, the watershed outlet (Fig. 5.3). A hydrograph represents the integrated response of all flow pathways from a watershed into the stream channel and includes the flow from all tributary channels in the watershed. The hydrograph, therefore, also reflects stream channel characteristics of upper channels through the main channel to the watershed outlet. As such, it is a principal tool used by hydrologists for describing and analysing floods and other streamflow responses.

The hydrograph shown in Fig. 5.3 illustrates the streamflow response from a rainstorm. Rainfall intensity is plotted in the upper part of the graph with the same timescale as streamflow to help visualize the time response of streamflow to the rain event. For the watershed depicted by the hydrograph, there is a relatively constant flow that precedes the rainstorm, called the *baseflow*. Baseflow, the result of groundwater flow into the channel, sustains the flow in perennial streams. The rapidly rising part of the hydrograph following the beginning of rainfall is the *rising limb*. The maximum flow rate is the *peak flow*, after which the flow diminishes. The falling limb of the hydrograph, called *recession flow*, represents the water draining from the watershed soils and tributary channels after rainfall ceases. Unless a subsequent storm follows quickly, the flow will eventually recede to near the pre-stormflow conditions or the baseflow. The total volume of streamflow that is above baseflow is *stormflow* and represents the quick flow response of the watershed to that particular rain event. There is only the stormflow hydrograph for intermittent or ephemeral streams as there is no baseflow component.

The quickest response that would be observed in the stream channel to a rainstorm is the rain that falls directly on the water surface in the stream channel, called *channel interception*. Channel interception causes the initial rise in a streamflow hydrograph. Wet soil areas immediately adjacent to the channel and overland flow immediately adjacent to channels quickly adds flow to the channel and, therefore, sustain the rising limb. Impervious areas and areas with soils that have infiltration capacities lower than the rainfall intensity would contribute flow quickly to stream channels. Some overland flow can be detained by the roughness of the soil surface, but it nevertheless represents a *quick flow* response to rainfall. If there are areas of higher infiltration capacities on the hillslopes, some of the surface runoff might infiltrate and become subsurface flow. Subsurface flow, also called *interflow*, represents the portion of rainfall excess – that is, the infiltrated water in excess of available soil water storage that travels within the soil and arrives in the stream channel over a short enough time to be considered part of the stormflow. There can be *zones of infiltration* and *zones of exfiltration* on a watershed with a mosaic of vegetation types and land-use conditions, where subsurface water emerges to the surface because of topographic change (toe of a hillslope) or because of change in depth of impeding soil layers.

The baseflow that occurs between rain events is usually maintained by groundwater flow into the channel. Groundwater flow does not respond quickly to rainfall because of the long pathways of flow involved and the slow movement of groundwater. Nevertheless, baseflow can change gradually over time and, during long dry periods, can become a small fraction of the baseflow that follows long periods of rainfall or snowmelt. The minimum flows that occur in a stream are important because they can dramatically affect water quality and aquatic life.

Prior to the initiation of hillslope studies, stormflow and floods were considered to be caused largely by overland flow and, therefore, assumed to occur only when infiltration capacities of watersheds were exceeded by rainfall or snowmelt rates. Stormflow was thought to be the sum of channel interception and overland flow with subsurface flow playing little or no role. With the advent of hillslope studies, however, it became apparent that subsurface flow plays an important role in generating stormflow from many watersheds. Subsurface flow has been shown to be the dominant pathway of flow on most forested hillslopes (Hewlett and Troendle, 1975; Brooks *et al.*, 2003; Chang, 2003). Furthermore, stormflow events from upland watersheds are the result of a dynamic process in which the source areas that contribute water to stream channels change with the antecedent conditions of the watershed. As more rainfall occurs on watersheds, areas of saturation at the toe of slopes expand – as do the channels that receive subsurface flow from hillslopes. This mechanism by which streamflow is generated is commonly observed in forested watersheds with deep soils and, importantly, helps to explain how streamflow following rainstorms or snowmelt can generate peak flows where surface runoff is negligible. This process is known as the *variable source area* concept (model) of the streamflow process.

Watersheds in arid regions often have lower infiltration capacities and shallower soils with lower soil moisture storage capacities than watersheds in humid regions. Surface runoff, therefore, is an important pathway of flow from watersheds. Watersheds in these regions tend to respond more quickly to precipitation inputs than watersheds in more humid (wetter) regions, with relatively higher peak streamflows for a given amount of rainfall excess than watersheds in other regions.

By contrasting hydrographs associated with different land-use conditions on a watershed and a given unit of rainfall, we can gain insight into how land use can modify stormflow behaviour. Land-use practices that cause more surface flow at the expense of subsurface flow will result in higher peak flows that occur more quickly. If a forest cover is removed and replaced with pastures and annual crops over a large percentage of a watershed, for example, both stormflow volume and the magnitude of the peak could increase.

Stream Channels, Floods, Flood Plains and Land Use

Stream channels on a landscape evolve in response to climatic, hydrologic and watershed conditions. Mountainous headwater streams in steep terrain cut deeply through soils and parent material, becoming entrenched in narrow valleys. The channel gradient is reduced as the terrain moderates downstream. As a consequence, the velocity of flow diminishes and the sediment that was removed upstream becomes deposited. These deposits form flood plains in the valleys over geologic time and the banks of stream channels become established in response to the higher flows that occur frequently on the flood plain. The *effective discharge* is the flow that does the most work in moving sediment over long periods of time (Rosgen, 1994, 1996). It is generally associated with the bankful flow and its high velocity; this flow occurs at least once a year on average. Floods occur when streamflow exceeds bankful conditions. Each time a flood occurs, the water spills over its banks and the over-bank flow velocity is reduced by the resistance of vegetation growing on the flood plain. The flowing water drops much of the sediment that had been carried by the faster moving channel flow as a result. This process continues to build the flood plain, which also happens to create areas attractive for growing crops and pastures because of their rich soils and favourable moisture conditions.

Over time, extreme rainstorms of high intensity and long duration cause the stream-flow peak to far exceed its banks, sometimes cutting new sections of channels and washing away sections of adjacent land. As flood peaks become larger, the area of inundated land expands (Fig. 5.4). The frequency by which floods reach twice the height of bankful is often associated with the 50-year recurrence interval (RI) flood, that is, a probability of 2% ($p = 0.02$) of being equaled or exceeded in any given year. The RI is the reciprocal of the probability (p) of occurrence in any year, $p = 1/RI$. The 100-year flood is often used to delineate flood-prone areas where constraints could be placed on human development activities. In contrast, the bankful flow is typically associated with an RI of about 1.5–2 years. The point we make here is that a *flood* is any flow that exceeds its banks and, therefore, represents a range of flows from the average annual peak flow (2-year RI) to flows that are associated with RIs of 100, 500, or more years.

A question that has long been debated is the extent to which land use and associated vegetative changes can increase flooding. There is no simple answer to this

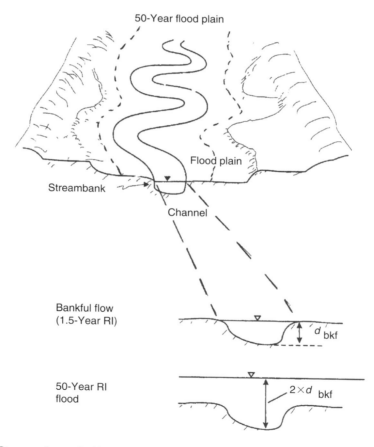

Fig. 5.4. Stream channel with corresponding bankful flow (bkf) and the 50-year recurrence interval (RI) flow on a flood plain. (Adapted from Brooks *et al.*, 2003.)

question. While land use can increase the magnitude of flood peaks associated with a 2-, 5- or 10-year RI in some instances, fundamental hydrology suggests that land use has a limited effect on major floods. More simply put, no matter how deep the soils, how many lakes and wetlands, or how dense the vegetation, the water storage capacity of any watershed will eventually be exceeded if excessive rainfall continues indefinitely. Unfortunately, there is currently no good understanding of the maximum flood threshold levels at which land use no longer has an affect.

Annual Water Yield

The annual water yield from a watershed is the residual of annual precipitation and ET as we have seen from annual water budgets. People in many arid regions of the world that experience water scarcity seek ways by which water supplies can be increased. Many options to achieve additional water exist, but they can be limited by geography, technology, economics and/or existing laws, regulations and policies. Conversely, some areas experience waterlogged soils that restrict some land-use practices. In such cases, there is interest in reducing water yield or flow at certain locations on the landscape. From a watershed hydrology perspective, opportunities to increase water yield focus mainly on modifying vegetative cover and soil conditions that either reduce ET (Bosch and Hewlett, 1982) or otherwise increase surface and/or subsurface flow.

Water Quality

The physical, chemical and biological characteristics of water flowing from watersheds, which cumulatively represent *water quality*, are determined by climatic and watershed conditions. Physical characteristics include temperature, dissolved oxygen, biochemical oxygen demand, suspended sediment and turbidity. The chemical composition of natural waters is the product of atmospheric deposition, natural processes of geologic weathering, soil erosion, solution or precipitation occurring in subsurface water, and, on many watersheds, human-induced chemicals (Hem, 1992). Biological characteristics can be considered from both a bacteriological and macroorganism or habitat perspective. Bacteria and protozoa are major concerns worldwide as their presence in drinking water affects human health. Macroorganisms range from macroinvertebrates to fish and serve as excellent indicators of overall aquatic health. Generally, the greater the diversity of species in a body of water, the better the overall water quality. As water becomes impaired, species diversity generally declines and we see limited species which have adaptations that allow them to survive in impaired water.

Water quality management objectives

When describing the collective, overall quality of surface water bodies (streams, rivers, lakes, etc.) or groundwater, we must indicate its intended use. Water used for drinking should meet more stringent quality standards than water that is to be used for irrigation. The establishment of total maximum daily loads (TMDLs) in the USA targets water bodies according to use, identifies where water does not meet these standards and requires states to take necessary actions to reduce loading to meet the standards. The TMDL approach, however, does not properly account for watershed

Box 5.2. Meeting total maximum daily loads: a point source approach to solve a non-point source watershed problem. (From Magner, 2006.)

Achieving improved water quality, and particularly the process of establishing a TMDL standard for natural water bodies, is problematic for several reasons. In the case of point source water pollution, meeting the standard is both a technological and economic issue where a wastewater treatment plant is required to meet discharge standards for different constituents. The water leaving a site through a pipe can be monitored and actions taken if standards are not met. With natural water bodies that are the product of watersheds, one question requiring an answer is whether we have the ability to detect if loadings are within the range of 'natural water quality', baseline or background levels. This is a more difficult issue than that of establishing standards for wastewater treatment plants or a drinking water facility.

The water quality characteristics of streams in a watershed are determined by the watershed's unique setting. Watersheds in differing locations often have varying background levels of nutrients, dissolved oxygen and so forth that can be due to natural processes – such as geologic weathering or the occurrence of wetlands with oxygen-starved water. Human activities in many river basin have so altered watersheds that it is difficult, if not impossible, to determine natural background levels of any constituent. Additionally, background levels of many constituents in undisturbed watersheds are not always constant and can vary with season and levels of streamflow discharge.

Suspended sediment and turbidity generally increase with increasing discharge that can vary widely during snowmelt runoff or rainy season flows. Turbidity can also be high during low flow periods, but this might be due to biological matter (such as algae) that is suspended in water rather than suspended sediment. The TMDL approach is flawed in that it ignores the natural processes of watersheds and, as a consequence, presents a dilemma to the agencies responsible for improving water quality and users of the land who are being asked to respond. With better monitoring and improved knowledge of watershed – water quality processes adjustments in the TMDL approach will likely be necessary.

processes and sets up conditions that are problematic for setting the standards and for those who are being asked to meet those standards (Box 5.2).

Water quality concerns

Most pristine watersheds, and particularly forested watersheds, have characteristically good water quality for many types of uses. That is, streamflow from pristine watersheds would normally have low nutrient content, low sediment, greater water clarity and higher oxygen levels in contrast to watersheds that have undergone intensive and often inappropriate land use. We generally consider watersheds that have high quality water to be *healthy* watersheds. Much like human health is attributed to body temperature, oxygen levels in the blood and chemical constituents in the body and blood, the same types of indicators can be used in assessing watershed health. Major water quality concerns, as they are both impacted by – and as they can be mitigated by – land-use activities, are discussed below.

When the vegetative and soil conditions of a watershed become degraded, we generally observe impairment of water quality. When riparian areas adjacent to stream channels or lakeshores become degraded, there is often a direct impact on water quality. At any time natural (undisturbed) forests or pastures are converted to intensive agricultural crop production, there is good potential for greater runoff, sediment and nutrient loading to streams or lakes.

Impaired water has direct and indirect effects on people living in upland watersheds and those living in downstream areas. Examples of water quality characteristics that can be affected by land use and (in turn) can impact downstream entities include (Lamb, 1985; Verry, 1986; Verry et al., 2000; Brooks et al., 2003):

- Suspended sediment in streamflow occurs naturally and increases with higher streamflow discharge. When land use accelerates surface soil erosion, gully erosion or soil mass movement along stream channels, levels of suspended sediment can exceed natural occurring levels that can adversely affect aquatic habitat and downstream use of water (hydroelectric turbines, reservoir capacity, etc.). High levels of suspended sediment increase turbidity which reduces light penetration in water and, therefore, can adversely affect aquatic life.
- Accelerated nutrient discharge into water bodies, such as nitrogen and phosphorus, can impair water through eutrophication. This process of excess nutrients causes algae blooms which can deplete oxygen levels.
- Temperature and dissolved oxygen can increase and decrease, respectively, when riparian forest cover is removed adjacent to small streams. Water temperature can increase to levels that impair aquatic life, directly, and can reduce oxygen levels in the stream as well. Cold-water fish and diversity of invertebrates can decline as a result.
- Removal of understorey vegetation and/or grass cover in adjacent channels can result in accelerated sediment and nutrient transport into the stream; it also causes increase in turbidity.

Recent decades have brought to light concerns about cumulative effects, particularly with respect to these effects on aquatic ecosystems and their health. Excessive loading of nutrients caused largely by non-point pollution from agricultural activities has caused widespread algae blooms, oxygen depletion and loss of sea life, affecting downstream fisheries in the Gulf of Mexico and several other estuaries.

Importance of Monitoring

Methods of measuring and monitoring the various hydrologic inputs and outputs and the various processes are explained in many hydrologic and watershed management references. Our discussion in this chapter emphasizes the importance of developing and sustaining climatic and hydrologic monitoring, which is crucial for the planning, implementation and evaluation of natural resource, agricultural, urban and other types of development.

Precipitation

The importance of long-term measurements of precipitation inputs to a watershed is emphasized in terms of understanding the hydrology of watersheds and in the context of planning for forestry practices, livestock grazing or agricultural cultivation. Long-term averages of precipitation have little value when there is extreme variability in year-to-year measurements. Precipitation patterns need to be understood better so that contingency planning can take place to deal with the severe effects of excessive precipitation leading to flooding or periods of low precipitation and prolonged droughts. Given the uncertainty associated with precipitation patterns in the future, monitoring is becoming more important for planning and considering future options.

Streamflow

Streamflow can be directly measured in most channels and, similar to precipitation, the longer the records, the better is our information for planning and management. Pre-calibrated structures such as *weirs* or *flumes* can be used to determine streamflow discharge directly from empirical equations for small watersheds studied for experimental purposes. Streamflow in large streams and rivers is most commonly measured by continuously monitoring the height of the water in the stream referred to as the *stage*. The stage is then converted to streamflow discharge by measuring the streamflow velocity and developing *stage–discharge relationships* that allow future stages to be converted into discharge.

Water quality

Monitoring water quality is important because of continuing concerns about sanitary drinking water supplies and uses of water for agriculture, industry and recreation. A watershed perspective is needed in establishing monitoring protocols so that adequate sampling of the necessary water quality constituents is conducted at the proper locations and within the appropriate time frame to be meaningful. Water quality monitoring has been haphazard at best from a worldwide perspective. Given that the analysis of water quality constituents is expensive, there is a need to improve upon monitoring programmes and water quality studies designed and targeted to address the most important questions in a particular area. The potential benefits to the health and welfare of people and the economic implications of detecting and treating impaired waters merit much greater attention to water quality monitoring programmes.

6 Monitoring and Evaluation to Improve Performance

Watershed management practices, projects and programmes that contribute to sustainable flows of natural resources generate activities that use, affect and are affected by the environment (Box 6.1). The condition of the environment (in turn) affects the economic and social activities and, ultimately, the well-being of people. It is a dynamic process where what we do one day can affect people in many ways in the future. Often, the purpose of implementing watershed management activities is to change land-use practices to improve the welfare of present and future generations without adversely affecting the environment.

The linkages among watershed management, environmental systems and the welfare of people are not always well understood. Although we plan for particular resource and welfare effects when a watershed management activity is undertaken, we cannot always be sure that the activity will be carried out as planned or that the activity will have the effects on natural resources, the environment and welfare of people as anticipated. This is mainly because our anticipations are based on imperfect knowledge and uncertainty regarding the knowledge that we think we have in hand. Consequently, establishing monitoring and evaluation (M&E) systems is a key element in the planning and implementation of land and water-use activities within the overall framework of IWM. What is unique about M&E in the context of IWM is that we must consider both the on-site consequences of practices on the land and the downstream, downslope or collectively off-site effects due to the practice(s). Without M&E, we have little basis to determine the effects of land and water-use activities on the natural resource base, the environment and the welfare of people. Similarly, we need M&E to facilitate changes

---

Box 6.1. Watershed practices, projects and programmes.

A watershed management *practice* is a change in the use of land, water and/or other natural resources and/or other actions that are taken by people to satisfy a land-use and/or a watershed management goal and a set of objectives. A watershed management *project* is a grouping of watershed practices, while a watershed management *programme* is a group of watershed management projects. People are rightly concerned with all three levels of planning and implementation. The main emphasis in this chapter is placed on *practices*, largely because we consider a practice to be the fundamental units, or building blocks in projects and programmes that most easily, clearly and concretely can be monitored in the planning and implementation of watershed management activities. Aggregating the monitoring of practices with a broader monitoring of the interaction of such practices leads to the monitoring of whole projects and programmes. At the level of practice, the evaluation that follows tends to be technical, while evaluation further up the scale adds increasing elements of policy evaluation.

---

in policies, practices and other activities that are required to meet desired goals, for example, as expressed in official regulations. Finally, M&E also provide the feedback that can help improve performance of future watershed management practices. They have a strong learning function when looked at in a dynamic context.

Relationship of Monitoring to Evaluation

Monitoring is the systematic process of collecting information to provide a basis for adjusting or otherwise modifying a watershed management practice that has already been implemented, identifying maintenance procedures for continuation of the practice and improving future efforts. Monitoring should begin at the start of a practice, project or programme and continue throughout its duration. *Evaluation* is the process of appraising the results of a managerial decision through knowledge obtained from both qualitative and quantitative monitoring activities. Evaluation is crucial in situations where there is little experience. Monitoring, therefore, is the process of collecting information and evaluation is the process of analysing this information to determine the worthiness of the effort implemented.

Managers need information about the activities of the management practice including input use and efficiency, outputs and their effectiveness, results over time and impacts. They also need early warnings of potential problems on the performance of the practice that might lead to the failure of the practice in meeting its original goals and objectives, which in turn can affect the larger aggregation of practices in a project. These potential problems must be known in time to make the necessary corrective actions. Planners and mangers require information about the performance of an ongoing practice to revise plans for the practices and improve the planning process for future practices. Administrators and managers need information to insure present and future practice accountability. Policy makers depend on M&E for information to develop more effective policies in relation to watershed management activities and the objectives behind them.

Monitoring

Monitoring is the gathering of feedback information about practice components, processes, activities and outputs as illustrated in Fig. 6.1. Monitoring can be a one-time activity to record accomplishments of the practice, for example, recording the number of surviving trees planted in a specified area, or the length of road built to an established standard in a specified time and what unintended or unexpected changes have resulted due to the road. Or, monitoring can be an activity where observations are taken at predetermined time intervals such as hourly measurements of the water flow in streams, monthly or annual amounts of sediment deposition in a reservoir or annual accountings of incomes of local people associated with a specific practice.

The monitoring of ongoing practices provides information that is essential for evaluating the effectiveness and efficiency of practices and management in general. Information on the time expended in completing specified tasks, amounts of materials used, records of financial expenditures and personnel performances, as well as accountings of individual and group accomplishments is helpful in scheduling work activities, evaluating work performance and providing records of accountability to administrators. Monitoring of outcomes (results) attained such as the number and

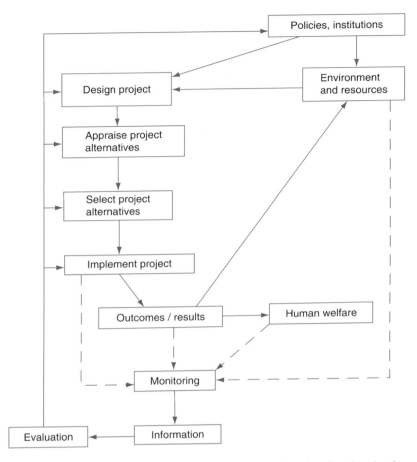

Fig. 6.1. Monitoring and evaluation provide feedback information for planning, implementing and evaluating the performance of future practices. (Adapted from Brooks *et al.*, 1990.)

cost of check dams built, the resulting decrease in sedimentation or increases in agricultural crop yields due to the implementing a practice provides information that is essential for operational planning and subsequent evaluation of practices. Monitoring of the welfare of people helps to determine the magnitudes of gains and losses in their health, education and income. This information is needed to document accomplishments and evaluate the benefits, costs and social impacts of the watershed management practice. Monitoring of the changes in the quantity and quality of natural resources and/or the environmental quality furnishes additional information about the impact of a practice on sustainability of its activities.

Evaluation

The determination of what needs to be monitored and how – what indicators are relevant, what data need to be collected and at what levels – depends directly on the evaluation objectives. Observations obtained through the monitoring of the components of a watershed management project provide the necessary information to evaluate and assess the past, present and/or anticipated future conditions and accomplishments, the emerging constraints and opportunities and the relationship between expected and actual performance. Such evaluation information is required to meet the needs of planners, managers, policy makers and other stakeholders. However, actual observations are rarely in a form that is particularly useful to people. These observations usually need to be brought together and evaluated in some way before they have value and can be used. Monitoring produces data, evaluation produces information from data. Evaluation, therefore, is the process of organizing and analysing the information data obtained in monitoring to estimate the value of past activities. Information is needed in order to make informed decisions about future activities. Evaluations also estimate the value of past accomplishments for an ongoing practice. They provide information about achievements attributed to the practice in the form needed by planners, managers, administrators and policy makers.

The terms *evaluation* and *appraisal* are often used interchangeably by people in referring to establishing and comparing the benefits and costs of a practice, project or programme. However, evaluation generally refers to impacts of past activities, while appraisal generally refers to expected impacts of present or planned future activities. Both are a form of *assessment*. Using this terminology, one would say that an appraisal is made to estimate the likely value of present or future activities such as weighing the expected benefits and costs of alternatives considered in planning of a watershed management activity (see Chapter 4).

M&E are often part of an iterative and dynamic process of adaptive management. M&E should take place throughout the planning and implementation stages of the practice. They provide feedback to the planners and managers to enable them to make changes in their plans and implementation activities as they go along. One cannot always specify the frequency needed for M&E because that will depend largely upon the type of monitoring to be undertaken, technology used and resources available. It will also depend on the rate of change in the external environment, which again cannot be predicted with any certainty. For example, weather information might be recorded daily, sediment accumulations on the bottom of a reservoir might be measured annually and data on human populations and other census information might be collected once every 5 or 10 years. A key here is that it costs money to collect information, so the monitoring process and the frequency and intensity of data collection should be designed to only collect information that is needed for a specific evaluation purpose; and that purpose should be justified within the broader context of the IWM programme and its requirements. M&E, therefore, provide feedback that is useful in guiding direction as the practices within an IWM framework evolve. If relevant and useful information is not being provided, then M&E objectives and methods should be modified accordingly.

Design of Monitoring Programmes

It is the responsibility of a manager to achieve the goals and satisfy the objectives specified in an agreed upon plan. The manager is accountable for expenditures and achievements. To be able to properly schedule practice activities, manage people and funds effectively and take corrective actions when necessary to change practice directions are essential. The manager must have information about activities, inputs and outputs associated with all the practices for which the manager is responsible. Three forms of monitoring are often considered in designing a monitoring programme for obtaining this information. They are:

1. *Implementation monitoring* gathers information to answer the general question: 'Did we do what we said we would do in the plan?' Here, we seek to monitor tasks and performance to document actual accomplishments. This form of monitoring might rely largely upon records kept by the personnel responsible for implementation, but it should also include some field checking to compare reports of activities accomplished with on-the-ground observations.

2. *Performance monitoring* gathers information to answer the question: 'Did the practice accomplish what it was intended to accomplish?' Specific elements to be monitored are determined largely by the goals and objectives of the practice. Here, changes in natural resources, environmental conditions and socio-economic conditions brought about by the practice are monitored. In the context of sustainable development, this question might not be asked immediately after the practice ends, because some of the project benefits will likely accrue beyond the practice's termination.

3. *Validation monitoring* gathers information needed to check on the assumptions made in the plan to answer the question: 'Were the assumptions and information used in the plan of the practice correct and did we choose the right practice given the assumptions?' Variables to be monitored might be chosen to check on particular data sets and assumptions used in planning and managing that are thought to be uncertain but play a key role in the practice outcome. This kind of monitoring is of value in all situations, but particularly when the results of the other two kinds of monitoring show inadequate performance or implementation.

Monitoring of all three of these forms is useful for managers and can be integrated closely into management plans to keep close control on the information to be provided and the cost of doing so. The information collected must be processed, evaluated and presented to the planners and managers in a usable format and timely manner. Monitoring has little value if it provides wrong information or information at the wrong time.

Determining information needs

Monitoring programmes should be designed to collect only the information needed by users for their particular evaluation purposes. Collecting, processing and storing data and then processing it to generate useful information is expensive. Only the data actually needed should be collected at a time and in a form that meets the needs. One of the first considerations in designing a monitoring programme, therefore, is to decide whose needs are to be met and what those needs are. Monitoring programmes must be linked closely to the people they are designed to serve. They should be highly interactive between the people who have knowledge about monitoring technologies and the people whose informational needs are to be met.

A useful way to hone in on essential information is to develop a clear statement of impacts sought from a practice and then trace it out in an *impact pathway* analysis on how it is intended to move from practice to impact; identifying critical information needs to verify how the practice in fact is performing along the pathway. For example, one practice might involve terracing. The desired and planned impacts are reduced erosion damage downstream and increased income or home consumption for the farmers using the terraced land. The impact pathway analysis traces out how the development of terraces leads eventually to the desired impacts. Thus, in terms of the goal of increases in farmer income, data need to be collected on the *without* terrace production and income flows and the *with* terrace estimated production and income flows, taking into account the likely different declines, if any, in productivity over time with and without the terraces. The same types of *with* and *without* data need to be collected over time on the additional costs associated with terracing and growing outputs on the terraces. With this framework in mind, the minimum data set to be generated by the monitoring activity can be identified.

Importantly, the potential benefits and costs of meeting different information needs must be considered in designing and implementing monitoring programmes. For example, if it costs more to generate daily rather than weekly streamflow data, then the value of the additional information should at least equal the cost of generating it. Therefore, for each monitoring activity, we should be asking: 'Is the value of the information at least equal to the cost of collecting it?' The value of information is not always easy to predict, but at least some qualitative judgement as to why it might be valuable needs to be made to justify the cost of the monitoring.

The objectives, purposes and inputs and outputs of a monitoring programme should be identified early in the planning stage for the IWM practices being designed to meet the project or programme objectives. Decisions must be made as to how the programme is to be administered and who will do the actual monitoring. It must also be determined whether the people doing the monitoring will be the same as those using the information obtained. If this is the case, it helps to insure that the information obtained will meet the needs of all users. One possible danger with this approach, however, is that the monitoring programme can become a routine collection of data for the sake of carrying out the management activity and, in doing so, fail to provide the information needed to evaluate all of the various effects. It is seldom that the given information is needed to evaluate only one element in a project or programme. Thus, it is important that we start with the evaluation needs and work our way back, for each need, to the specific data requirements, identifying common data needs in the process.

Alternatively, monitoring might be carried out by an organization that is independent of the users of the information. This latter approach can be the most practical if monitoring serves a number of groups and evaluation needs. Also, in cases where the M&E is done for accountability or due diligence purposes, independent groups often are desirable since they provide the arms length assurance that the results are credible.

A potential problem with both approaches is that those responsible for monitoring can become so involved and caught up in the process of collecting data and information that they no longer consider appropriately the needs of specific users. Monitoring becomes an end in and of itself, a solution in search of a problem rather than the other way around. This is less likely to happen with the independent monitoring if it is the user of the results – the project or programme manager, planner or administrator – who controls the contracts for the independent monitors. Past experience indicates that large volumes of data that are of little use are often collected in monitoring efforts. Monitoring activities can also be split so that some activities are performed by the users and others by a group that is independent of the users.

The users to be served by a monitoring programme must be consulted early in the process of designing the monitoring programme to determine their information requirements. Those designing a monitoring programme should not assume that the users know and/or understand the specific data needs required for an evaluation purpose. Often, the managers are generalists who understand management well, but they do not have all the technical knowledge to design specific M&E functions. Ignoring the need to work with managers to identify true information needs results in the risk of designing a monitoring programme and collecting information that will not be used. Many file drawers and computer files are full of sets of data and other types of information that were stored in the expectation that the data and information would be useful in the future, but they were never used subsequently because they failed to meet the actual needs of the users.

Establishing what should be monitored can be challenging in that there can often be many objectives and questions to be addressed. Some monitoring is routine in nature in that it accompanies the normal ways of doing business. For example, maintaining or improving on-site productivity and sustaining that productivity through environmentally sound land-use management is – or should be – inherent to all watershed management practices. Monitoring on-site productivity can be accomplished through gathering information on the levels of productivity associated with the agricultural crop, livestock, and/or forest production activities over time. It is safe to assume that most land users are interested in year-to-year production and, therefore, might have such records on hand. However, in many cases, long-term records of on-site productivity are not readily available and there is the problem of determining on-site productivity when land uses change over time. Declining production levels over a long period of time could also be symptomatic of changing climatic conditions or other factors that are outside the control of the watershed on-site productivity activities associated with the watershed management programme. Analysis of local weather records would help determine rainfall or temperature trends, while loss of soil fertility might be reflected by periodically inventorying soil characteristics and properties. There can also be other effects to consider including insect and disease outbreaks, fire and other anomalies that must be taken into account. What should also be incorporated into any monitoring programme of a watershed management practice are the hydrologic and other environmental conditions that affect downstream (off-site) interests. Finally, there is the fact mentioned earlier that the objective of many watershed management activities is not to increase or maintain on-site productivity but rather to reduce the loss of such productivity. This requires setting up methods of estimating what the productivity loss would have been without the practice, project or programme.

Monitoring can also provide information about changes in land use, precipitation and temperature associated with climate variation and change (Box 6.2). Understanding how the amount and quality of water flow from watersheds have changed over long periods can be of great importance to plan for the future. As seen in Box 6.2, by resurrecting old climatic and hydrologic monitoring stations, often abandoned because of economic reasons, it is possible to determine long-term trends, and other natural changes on the land that affect both watershed production and water flow.

Box 6.2. Reinstating hydrologic monitoring stations to acquire information about long-term effects of land use and other effects on hydrologic processes.

Long-term monitoring can be invaluable in improving our understanding of watershed effects of land use, help quantify watershed response to extreme events such as floods and droughts and to better assess changes in hydrologic response over time. The natural variability of responses can be gained through continued monitoring of experimental watersheds that have been used as controls (Stednick *et al.*, 2004). In some instances abandoned watershed studies might be reinstated as barometer watersheds that can extend the records of previous decades. Similarly, monitoring on major rivers and watersheds that has been discontinued for various reasons (in the USA, the US Geological Survey (USGS) and the National Weather Service are responsible for maintaining long-term streamflow records and climatic records, respectively) can be reinstated to consider questions related to watershed response to insect infestations, fires, droughts and extreme rainfall and/or snowmelt runoff events. Changes in watershed conditions may be detected by monitoring watersheds at different scales (nested watersheds) across watersheds that have undergone land-use change. In addition to USDA Forest Service watersheds, the Agricultural Research Service (ARS) and Bureau of Land Management (BLM), experimental watersheds can also be used to investigate a greater range of land-use conversions. USGS gages on basins with expanding urbanization and forest change can likewise be used to monitor annual water yield and stormflow change associated with different recurrence intervals.

Resurrection of watershed studies and monitoring stations that have been abandoned over the past half century offer unique opportunities that await measurement and analysis. As examples:

- Re-establishing monitoring of the Wagon Wheel Gap watersheds in Colorado (REF) could allow comparisons of the old data on forest cover change with the ranching and second home development that now exists on the watershed.
- The Watts Branch in Maryland, where Leopold *et al.* (1964) conducted their fundamental studies of channel-forming flow, was abandoned by the USGS and has subsequently undergone extensive land-use change and down cutting of the channel. Re-establishing monitoring on such sites would allow comparisons of flow regimes under current land-use conditions with those of the 1950s.
- Similarly, old monitoring stations and experimental watersheds that have been closed down, and that subsequently had fires and/or insect infestations could be resurrected. Small instrumented watersheds in the ponderosa pine forests of central Arizona were reinstated following the intense and highly destructive Rodeo-Chediski fire of 2002, providing excellent post-fire effects that can be compared with pre-fire conditions of the 1970s (Ffolliott and Neary, 2003). For example, these authors found that the intensive fire that nearly complete elimination of the forest overstorey on sites where a high severity burn occurred resulted in significant reductions in interception losses. Extensive water repellency formed on severely burned areas, causing a decrease in infiltration and, as a consequence, increases in overland flows. While the reduction in infiltration decreased the amount of water entering the soil, the loss of vegetation to burning meant that less water was removed from the soil by transpiration. Changes in soil water storage coupled with the occurrence of a water-repellent

Continued

Box 6.2. *Continued*

soil contributed to the higher overland flows. Shortly after the wildfire, a peak flow measured on a severely burned watershed was about 2350 times the highest peak flow recorded before the fire. This peak flow was also the highest known post-fire peak flow in the ponderosa pine forests of the south-western USA. The sediment-laden water flowing from burned areas contained large amounts of organic debris, dissolved nutrients and other chemicals released by the wildfire. Some of the sediment- and organic-rich water flowed into the Salt River, the major tributary to the Theodore Roosevelt Reservoir, which is a main source of water for Phoenix and its surrounding metropolitan communities.

The costs and benefits of monitoring information

The above section suggests that we can often gain valuable information at a reasonable cost from monitoring data that are outside the original intent of the ongoing or newly planned monitoring activities focused on the needs of a new project or programme. Regardless of how we generate the needed information, economics dictates that we be prudent in establishing monitoring activities. The benefits from the information gained through monitoring must at least be equal to the costs of obtaining, analysing and providing it to users. A main goal of monitoring is not to provide the most detailed information possible, but rather effectively and efficiently meet the information needs of the users. A question to be asked to users is: 'How much information do you need to make a *reasonably* good decision?' Defining minimum required data sets is always a prudent way of starting the design of an M&E system. Obtaining too much information in too much detail can interfere with decision making rather than help it – particularly if it creates confusion among the managers who need the information most. Users often suffer from informational overload by having so much information available that they cannot digest it all and get confused on how to use it, therefore, wasting valuable managerial time on issues that are really not priorities.

In establishing information requirement priorities, the costs of obtaining, analysing, storing and retrieving and presenting the information must be compared to the potential benefits of having the information collected. Those designing a monitoring programme should recognize the time and costs of training the necessary personnel, obtaining the information, doing the evaluations and writing the reports. Before reporting requirements are imposed, however, the benefits of having the information and of collecting or obtaining it should be evaluated. Furthermore, monitoring and reporting requirements should be periodically reviewed to determine if the information initially deemed necessary is still needed, if the costs have changed for collecting it and if the information priorities should be modified. Determining information-generating priorities is an iterative and interactive process between the people who will provide the information and its users. By themselves, these two groups might have limited effectiveness in establishing the information priorities. Working together, they can be more effective in designing an optimum M&E process and framework for given projects and programmes.

Collecting information

Collection of necessary information for monitoring programmes requires many types of instruments and sampling networks which insures that the data to be collected make sense and are valid and useful. Details on instrumentation and commonly employed sampling techniques and monitoring protocols for a number of biophysical data topics are available from Conant *et al.* (1983), Kunkle *et al.* (1987), Ffolliott (1990) and Wiersma (2004). Appropriate sampling procedures are also critical to efficient collection of socio-economic data. Questionnaires or interviews with the people inhabiting the area to be impacted by the practice are common methods of obtaining information. Use of questionnaire and interview methods must be thought of carefully, however, so that the information obtained is useful and imposes minimum inconvenience on those being interviewed. For example, while estimating the income of local people, which can often be accomplished with questionnaires and/or examinations of tax records, the questionnaires can be subject to reporting errors because of a reluctance of the people to disclose their personal incomes to hide true tax obligations or because people think that personal information is of no business to anyone other than their employers and the tax people.

The task of collecting information can also be difficult because only a relatively few standard techniques are available for initiating a monitoring programme. This difficulty – when it occurs – is attributed largely to the fact that no single set of guidelines for collecting monitoring information apply in all situations. As a consequence, one might find that the best allies are one's own creative instincts and those of other people who are either cooperating with the practice to be monitored or have confronted similar problems in other areas. A number of approaches to collecting monitoring information can apply in some situations depending on the objectives of the practice and circumstances in which it was implemented. Nevertheless, one or more of the following categories of measurements are usually involved in collecting information for a monitoring programme:

- Direct measurements of inputs and outputs of a watershed management practice using the most appropriate mensurational techniques and sampling procedures;
- Indirect measurements (secondary data) obtained through solutions of predictive relationships between inputs and outputs when these relationships are available for the practice;
- Imputed measurements of changes in the sustainability of the practice and welfare of people both on and off the area impacted by the practice derived largely from indices of internal and external changes.

Collecting baseline data on water quality to compare with water-quality standards is a frequent goal of a monitoring programme that is designed to obtain information required to evaluate the effects of a watershed management practice on water-quality constituents. When an extensive monitoring programme is required, it often becomes necessary to depend on volunteers to work with experienced professionals in collecting the specified information by following established procedures to insure data credibility and consistency. One example of a generic procedure for this purpose is presented in Annex 6.1.

Information management

Information is expensive to collect, analyse, store and later retrieve. Therefore, considerable attention should be paid as to how the collected information is going to be managed. A description of what data sets are to be collected and how the data will be processed, stored and made available to the users should be part of the design of a monitoring programme. For example, information management with computerized data management programmes can be one component of the monitoring programme. Jane *et al.* (2004) emphasize the need for integrated data management for environmental monitoring programmes that adequately addresses multidisciplinary data across varying spatial and temporal scales. The value of the IWM approach is that we consider different spatial and temporal scales and the linkages that exist among land and water use in terms of uplands and downstream effects. Information management systems should be compatible with this approach, but they must also be reliable (quality control is essential), secure, flexible in responding to changing conditions, as well as accessible and compatible in providing timely access to information through computer networks and other means and be supportive of the types of analytical methods to be employed in the evaluation process. Moreover, the system must be cost effective in terms of providing the minimum data requirements of the IWM programme.

When it is necessary to collect large amounts of information and then store this information for subsequent retrieval and analysis, a database management system can become necessary. Although it is beyond the scope of this chapter to elaborate in detail the many kinds of database management systems that might be appropriate for a specified purpose, a general description of the characteristics of database management systems and a few of the database models are presented in Annex 6.2.

Monitoring the Biophysical System

The purpose of monitoring the biophysical system is to determine whether the activities in a watershed management practice, project or programme have had the effects anticipated as illustrated in Fig. 6.2. Importantly, the monitoring effort must have a specified purpose. If erosion control practices involving the planting of a cover of trees, shrubs or herbaceous plants are implemented to improve the upland productivity of a watershed and reduce downstream sedimentation, measurements of the consequent changes in soil erosion and sediment transport are needed, as well as the changes in water flows downstream, recognizing that tree cover can result in less water downstream. The changes in soil erosion must also be related to both upland productivity and downstream sedimentation. However, such comprehensive, watershed-focused M&E is seldom carried out, perhaps for cost reasons and perhaps because of lack of full understanding of what is involved in such comprehensive measurement.

To determine effects of a watershed management practice on biophysical conditions, it is necessary to monitor where effects of the practice can be measured. However, it is also desirable to monitor a control area so that a comparison can be made of areas with and without the practice as outlined in Fig. 6.2. This is in keeping with our earlier comment that the objective is often to reduce losses rather than

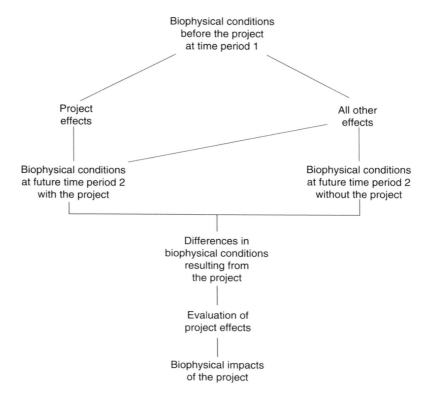

Fig. 6.2. Comparing the impacts of a watershed management practice on biophysical conditions. (Adapted from Olsen *et al.*, 1981.)

increase something. The 'with and without control' can be useful in this case. The control for such a comparison must consist of information obtained from the area to be impacted by the practice before it was actually implemented in some situations. Comparisons can then be made between the before and after conditions of the watershed area (see Chapter 4). However, these comparisons involve some risks.

Types of biophysical monitoring

Several types of biophysical monitoring of a watershed management practice can be carried out, each with its own purpose (Ponce, 1980; Ffolliott *et al.*, 1995a). Examples of these types of monitoring include:

1. *Cause-and-effect monitoring* is designed to quantify effects of specific activities in a practice on measurable outputs (see Figs 2.3–2.6 in Chapter 2). Cause-and-effect monitoring helps to answer questions of what happened and, in doing so, understand why it happened as it did. Examples of questions that might be asked include:

- Did a conversion from forest vegetation to grassland pastures on 50% of the upland watershed increase the supplies of water?

- Were on-site soil erosion rates reduced and, if so, were sediment deposition rates reduced at the reservoir site?
- Did the establishment of riparian forest buffers improve water quality?

2. *Baseline monitoring* provides planners and managers with reliable information on the variability of climatological, physical and biological data. This kind of information is useful at early stages of planning a future practice. Baseline information helps people answer questions such as:

- What are the variations in annual precipitation and length of drought periods?
- What is the extent of differences in soil fertility?
- What are the growth and annual yields of agricultural crops?

3. *Compliance monitoring* is used to determine if established standards and/or criteria are being met. This type of monitoring also provides information that is useful in the planning of future practices. Compliance monitoring is problematic when standards are established without proper understanding of the interactions between land-use and watershed processes. This issue has implications for watershed management mitigation of non-point source pollution (Box 6.3). Questions that might be answered include:

- Are farmers planting land for which they are receiving payment?
- Are limits to irrigation use being exceeded?
- Are drinking standards being met?

Box 6.3. Integrating sentinel watershed systems into monitoring and evaluations of water quality: a case study (From Magner, 2006).

Section 303(d) of the Clean Water Act in the USA requires that individual states identify bodies of water that do not meet water-quality standards and prepare a TMDL assessment for water bodies identified as impaired. Historically, the management of pollution in Minnesota has focused on point-source regulations. However, non-point source pollution from upland watersheds has become the 'largest driver' of impaired water bodies in the 21st century. Pollutants such as sediment and nutrient imbalances are often associated with poor land-use management practices, but the cause-and-effect relationships can be elusive because of natural watershed-system influences that vary with scale. Elucidation is difficult because the present water-quality standards in Minnesota were designed to 'work best' with water-quality permits aimed at controlling point sources of pollution, for example, the pollution from municipal waster-water treatment plants. An IWM approach has been suggested to better monitor and assess non-point pollution in streams, rivers and lakes. This approach integrates physical, chemical and biological data collected through space and time on selected watersheds to separate natural from anthropogenic causes of non-point pollution. The hydrologic processes and functions of wetlands, riparian ecosystem and stream channels are more directly taken into account. Long-term and state-of-the-art monitoring and assessment must accompany this IWM approach to, ultimately, improve Minnesota's water quality through the establishment of more realistic standards. Ecologically based water-quality standards that integrate physical, chemical and biological criteria offer the potential to better understand, manage and restore impaired water bodies.

Role of computer simulation modelling

To address questions about land-use effects on hydrologic response, computer simulation modelling not only complements monitoring data, but also provides a means of effectively utilizing monitoring data. Questions have been asked for many decades about such effects as the extensive drainage of wetlands, the channelling of riparian corridors and large-scale deforestation on the occurrence of flooding on a watershed or within a river basin. Some people have claimed that the conversions of forest cover to agricultural or urban landscapes have increased the magnitude and/or frequency of flooding. To address such questions from a monitoring standpoint might require 100 years of monitoring streamflow regimes for the varying land-use conditions being compared. One solution to this situation can be through predicting streamflow responses with hydrologic simulation models that represent the land-use conditions encountered. Long-term or stochastically derived precipitation data are often used as input to these models from which simulations of streamflow responses can be analysed to derive estimates of the magnitude of flood events associated with the 75-, 100-year and greater recurrence intervals.

A general discussion of the methods of development and applications and some of the criteria for selecting a computer simulation model for use in watershed management is found in Annex 6.3.

Evaluating cumulative effects of land-use practices on watersheds can also be achieved through computer simulation models that can take into account the required monitoring information collected over time and space and then relate such data to land-use changes with the application of a Geographic Information System (GIS). For example, the hydrologic affects of changing land-use practices on a watershed or within a river basin can often be examined for varying scenarios of climatic change through the use of GIS and computer simulation modelling applications (Box 6.4). Given the present interest in climatic conditions (temperature regimes, precipitation amounts and pattern, etc.) on land and water resources, such tools are invaluable. Formats for storing data layers in a GIS, some of the more common applications of a GIS and the potential sources of error in a GIS are considered in Annex 6.4.

Box 6.4. The use of computer simulation modelling and GIS technology in estimating soil erosion: one example of this combined application.

Coupling a computer simulation model of soil erosion with GIS technology is providing watershed managers in Arizona with needed information on the spatial impacts of wildfire on post-fire soil erosion rates on upstream watersheds in montane forests. The burned watersheds have been divided into 'modelling elements' largely on the basis of physiographic features (including slope position and percent, aspect and soil characteristics and properties) and vegetation (post-fire changes in cover), with each of the elements characterized according to the required modelling inputs and the data input into the computer simulation model. This simulation model is based largely on the Modified Universal Soil Loss Equation developed in the USA by the USDA Agricultural Research Service (Williams, 1975; Wischmeier, 1975; Clyde *et al.*, 1976; and others). While the simulation results vary with the physiographic location of the watershed and the severity of the wildfire, this coupling of methodologies has become a management tool of increasing value in the region.

Monitoring Socio-economic Impacts

Some watershed management practices seek to protect the natural resources that people use to satisfy their needs and enhance the welfare of people as a result. Other activities seek to directly change the way in which people use the natural resources available to them. Still other practices attempt to improve productivities of the natural resources in meeting people's needs. Therefore, either directly or indirectly, IWM practices are intended to affect people and their living standards. Because these practices are people-oriented, watershed managers must be concerned with monitoring changes in people's activities that are brought about by the practice. Managers, therefore, must incorporate the collection, analysis and evaluation of socio-economic data and other information into their monitoring programmes. Similar to biophysical information, socio-economic information must be collected so that differences between the with and without conditions imposed by the watershed management practice can be used to determine the impacts as shown in Fig. 6.3.

A means of obtaining the information necessary to establish the with and without socio-economic conditions must be found in achieving the above practices. As with biophysical monitoring, it is rarely possible to monitor a control area for an IWM practice that exactly matches the area impacted by the practice. It might be

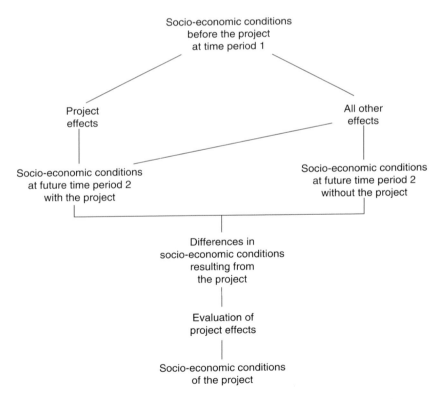

Fig. 6.3. Comparing the impacts of a watershed management practice on socio-economic conditions. (Adapted from Olsen *et al.*, 1981.)

feasible to compare before and after observations of societies being affected by the practice in some instances. But, it must be recognized that changes are likely to occur in a society even in the absence of the practice. Considerable judgement and experience is, therefore, required to interpret and evaluate observations obtained in monitoring socio-economic impacts. Most socio-economic monitoring in an IWM context starts with the biophysical information generated through monitoring, moves on to measure the people variables associated with biophysical changes and finally attaches values to such variables, where relevant. Socio-economic monitoring should not collect large amounts of data on a wide range of variables unless these data sets are required by law or regulation or needed by the planners, managers and policy makers for planning or accountability purposes. A key is identifying whom to ask to determine specific socio-economic informational needs. For example, if the activity or outputs to be monitored affect and are the responsibility of women and children, then women and children should be involved in determining the information needs. If subsistence farmers are beneficiaries or those who incur costs, then variables should be chosen that are relevant for the lives of such persons.

Design of M&E Systems

M&E systems are designed to meet the needs of the planners, managers and policy makers. It is essential, therefore, that evaluation procedures be based on the result of a dialogue between the evaluators and the users of the information to insure that pertinent and useful information is generated and provided to the relevant users in a timely fashion. If evaluations are not user-driven and designed to meet user needs, they are unlikely to be used as they should be. Evaluation systems should be independent of the activities being evaluated, however, to provide the necessary objectivity, transparency and credibility. In contrast to monitoring programmes, however, Casley and Kumar (1987) suggested that evaluations be undertaken from a central location – for example, in a state, regional or national office. Reasons for selecting a central location for the evaluation effort include:

- The high-level of skills needed for evaluation for every conceivable IWM practice might not be available on-site but can be available at central locations.
- Data and the other information required for evaluation – particularly information on long-term resource, environmental and socio-economic impacts – might need to be collected in a time span that is longer than the practice duration. This information is often stored at a central location.
- It might be necessary to gather information beyond the boundaries of the area impacted by the practice to make comparisons. Once again, a central location facilitates the efficient storage of such information.
- Standard methods of evaluation should be used to facilitate the comparisons of collected information for all the linked watershed management activities on a watershed or river basin. This might be difficult to achieve if each practice conducted its own evaluation and retained the results of the evaluation.
- An independent evaluation unit at a central location is likely to provide a more objective and credible evaluation of a practice than the personnel involved in the practice.

Determining information needs

Evaluation systems must be closely linked to monitoring; and the requirements of evaluations should be used to design the monitoring programme. However, evaluations usually require additional studies to provide the information not obtained through monitoring. For example, monitoring can provide information as to the *extent* that a particular IWM practice has been adopted by the local people for future application. But, monitoring is unlikely to explain *why* some people did not adopt the practice or why some people are early adopters and others are followers. A monitoring programme usually is confined to the more routine information related to what happened, while some evaluations seek to analyse a situation to determine what the human impacts have been and why something happened and suggest what might be done to correct an undesirable situation. This might require additional study to gather information not obtained by the monitoring programme.

Many of the tools and approaches discussed in Chapter 4 are relevant to ongoing evaluation needs even after the project or programme has been planned and implementation has started. IWM involves a dynamic and iterative process of generating successive approximations and making generally marginal adjustments towards some overall goal. Evaluations are concerned with extracting the relevant pieces of information from a mass of data sets and presenting this information in a form that is easy to use in moving on with this iterative process. To do effective evaluations – i.e. ones that are relevant and useful – requires an understanding and appreciation of the data sets and supporting information available, methods of analysis and informational needs of the user. Information has little value if it cannot be used or has been received at the wrong time. Therefore, attention should be given to the means by which the collected data are anticipated to be analysed and the information presented to its prospective users. An evaluation of a watershed management practice must provide a complete analysis of the performance, effectiveness, efficiency and impacts of the practice in relation to its stated objectives. Thus, evaluation at the practice, project or programme level attempts to:

- Critically re-examine the rationale that was stated in the planning documents in light of subsequent developments;
- Determine the adequacy of what is being done to overcome the constraints imposed on its execution and, as a consequence, promote the desired changes;
- Compare actual achievements with targets set initially and identify reasons for over-achievements and shortfalls;
- Assess efficiency of the procedures employed and quality of the managerial performance;
- Assess the economic efficiency associated with the practice, project or programme using standard benefit cost approaches (see Chapter 4);
- Determine the environmental impacts of what is being done or has been done;
- Present lessons learned and the recommendations that follow to future planners, managers and policy makers.

Many of these objectives and the methods associated with them are parallel to the ones discussed in Chapter 4 which dealt with planning of practices, projects and programmes *ex ante*. In fact, the *ex ante* and *ex post* evaluation or assessment activities should be linked so that there is a learning process for the next time an *ex ante* planning process is undertaken. We once again stress that we live in a dynamic world, where the key is to learn from the past.

Timing of evaluations

Evaluation of a watershed management practice can take place at varying times (stages) in the life of the practice (Casley and Kumar, 1987; Ffolliott *et al.*, 1995a). In terms of time, the following types of evaluation are common:

- Continuous and ongoing evaluation is often an informal type of evaluation that is conducted by the personnel involved to improve the planning and management of the ongoing practice. Results from this type of evaluation are used to achieve the goals and objectives of the practice better. This is the type of evaluation that indicates 'early warning signs' of potential problems in the conduction of the selected practice (see below).
- A midterm evaluation is carried out when the practice is approaching the midpoint of its duration to determine how well the implemented practice is functioning to this point. Performance of the practice is also evaluated to find out how specified components are performing, appraise their accomplishments and determine whether they are achieving their objectives on time. The results of this type of evaluation are useful in deciding whether major changes are needed in the practice.
- A termination evaluation is conducted at the end of the practice to document the overall activities and accomplishments for inclusion in the completion report for the practice. Here, the emphasis is placed on reporting what the practice achieved in relation to its original goals and objectives. Departures from these goals, objectives and other targets need to be explained and justified and unexpected consequences described. Attention should be given to the possibilities of sustaining the benefits of the practices after its termination, with recommendations as to how these benefits can be sustained. A termination evaluation is also intended to provide information to improve planning of the performance of future IWM practices. Therefore, unsuccessful components of the practice should be documented along with successful components, so that we do not repeat failures of the past.
- An *ex post* evaluation is conducted several years after the practice has been terminated to document its long-term effects, accomplishments and impacts. This evaluation can be important because the effects of many watershed management practices continue long after the practices terminate. For example, a protective vegetative cover planted to control soil erosion might provide lasting benefits for decades following its establishment. Similarly, it could take several years after a practice's termination to adequately evaluate the effectiveness of measures to control sedimentation. Therefore, long-term benefits and costs of the practice can be identified and quantified only through long-term evaluation. Unfortunately, *ex post* evaluations are rarely reported in the literature, which is surprising because it is well known that many of the benefits and costs attributed to watershed management activities are long term in nature.

Early warning signs

Evaluation that is carried out throughout the duration of an IWM practice provides feedback information to help in guiding the ongoing direction of the practice. Such continuous and ongoing evaluation will help to determine how things are going and provide early warning signs of potential problems in continuing the practice (Table 6.1). Corrective actions can often be taken if potential problems are known in time.

Evaluation focus

Evaluation should focus on different aspects of an IWM practice to be effective regardless of its type. Casley and Kumar (1987) mentioned three aspects that are particularly important – they are (i) performance; (ii) outputs, effects and impacts; and (iii) financial and economic efficiencies. All of the activities of a project or programme should be evaluated in terms of performance to determine the extent to which each contributed to achieving the practice goals and objectives. This focus includes a review of goals and objectives, implementation problems and managerial and financial performances. This evaluation focus can also help to improve the managerial performance by indicating how a manager should be evaluated in the future.

Evaluation should also focus on the outputs, outcomes, effects and impacts (refer to Figs 2.3–2.6 in Chapter 2). The activities of the project might have been completed successfully, but the results obtained might be different than those anticipated.

Table 6.1. Examples of early warning signs of potential problems encountered in carrying out a watershed management practice.

Category	Examples
Resource and environmental	Decreases in growth and yields of plant resources
	Adverse changes in plant species composition
	Lessening of livestock carrying capacities
	Loss of soil resources to erosion
	Adverse changes in water yields, timing of flow and water quality
	Loss of aquatic life
	Loss of biological diversity and ecological stability
	Loss of wildlife resource
Socio-economic	Adverse changes in employment and working conditions
	Detrimental impacts on local cultural traditions
	Increase in local and regional vulnerability to eternal pressures
	Loss of political stability
	Decreases in public participation
	Adverse allocations of resources
	Lowering of the levels of production
	Loss of stabilities in incomes

Suppose that an objective of the practice was to reforest a denuded watershed, while the target for the practice was expressed in terms of planting a specified number of hectares with trees. Tree planting could have been completed on time, but only a few hectares might have been *successfully* reforested if most of the trees subsequently died. It can, therefore, be seen that the target of a practice is often not stated in terms of the goals and objectives of the practice. An evaluation is needed to determine if the activities of the project actually accomplished the stated goals and objectives. An evaluation is also needed to determine what secondary, unplanned or unintended and external effects and impacts occurred and, if they did, whether they resulted from the activities of the practice. In economic terms, this amounts to an assessment of the externalities – both negative and positive.

A major focus of evaluation should also be placed on obtaining the information necessary for analysing the financial and economic impacts of the project or programme. Knowledge of these impacts are useful in planning for future watershed management practices to meet similar goals and/or objectives. Several measures of financial and economic efficiency are generally used for this purpose including distribution of benefits and costs, the present net worth, internal rate of return and benefit and cost comparisons (see Chapter 4).

7 Research, Training, Information and Technology Transfer

The main objective of this book has been to improve the ability of people to become more effectively and efficiently linked to their land, water and other natural resources through applying an IWM approach. However, the information available to achieve this improvement is insufficient in various ways. While the normal feedback mechanisms involved in adaptive management, and the lessons learned from the past can satisfy some of the needs, it often becomes necessary to undertake an investigation to obtain the additional, but unavailable information, i.e. to initiate a research effort. For research effort to be of value requires an understanding of the contributions that research can make in linking people to their land, water and other natural resources, the respective roles of watershed researchers and managers, as well as how research can be planned and conducted to meet the expressed needs of the people.

However, to have value, the information that is obtained through research efforts – along with knowledge gain from managerial experiences – must be made transferable to the stakeholders in a useful format. The mechanisms that are used for the transfer of this information and technology to people at all levels of interest include training and education in a classroom setting; distance learning to reach people who are separated by time and place; and, more recently, applications of Internet technologies including the World Wide Web (WWW). The contributions of research in watershed management and the mentioned approaches to transferring information and technology obtained from research findings and management experiences are discussed in this chapter. In the context of the IWM process, where participatory management involving all the various stakeholders is essential, these functions of appropriate research and effective knowledge sharing become particularly important. It is critical that all stakeholders share some common knowledge of objectives and means and that each stakeholder group has a minimum understanding of the specific information needed to carry out its functions within the broader IWM process.

Contributions of Research to Watershed Management

Science, as the term is broadly used in this chapter, represents the possession of knowledge attained through study (research) and practice (management). Therefore, both researchers and managers are scientists. In fact, there is blurring of their contributions to knowledge in what is called *action research* and learning by doing. It is our view that scientists – again, both researchers and managers – must enter into a compact of trust with other interested people in the advancement of knowledge on the management of watershed landscapes. Unfortunately, researchers and managers are occasionally at odds on the direction that watershed science should take. Much of this disagreement can be traced to differing philosophies that researchers and managers possess and the presumed differing roles they play.

Roles of researchers and managers

The roles of watershed researchers and managers in relation to the uses of natural resources in satisfying people's desires have been articulated by Stoltenberg *et al.* (1970) and Ffolliott *et al.* (1995b). In a general context, watershed managers[1] assist individual people and society as a whole in a number of ways. Watershed managers can help people to identify and clarify their goals and objectives. They also identify alternative approaches to achieving their goals and objectives and help and work closely with them in evaluating or comparing these alternatives, selecting the most promising alternative for achieving the stated goals and objectives. However, watershed managers often spend a disproportionate part of their time supervising activities necessary to implement these decisions and, as a consequence, have insufficient time in many instances to focus on problem-solving efforts. This situation is unfortunate, because watershed managers are often making their most valuable contributions helping people to make decisions.

Watershed managers' contributions are frequently measured by how effectively they can help people to satisfy their expressed wants, while the value of watershed researchers is determined largely by how much their investigative efforts increase the manager's efficiency to plan and implement watershed management practices. One purpose of watershed research is developing new concepts, tools and services for watershed managers. Another purpose of such research is answering questions that arise in the management process. A third purpose is answering questions that arise in conducting research itself since all too often these basic questions must be answered satisfactorily before the first two purposes of research can be efficiently addressed.

Some watershed researchers do not furnish their results directly to watershed managers. Rather, they provide information to solve the problems encountered by other, generally more applied researchers working more closely with the managers. Watershed research, therefore, can be viewed as a continuum with the watershed manager at one end of the spectrum. Ironically, along this continuum, it is common in all research circles to refer to *upstream research* as being the more basic research in such relevant disciplines as chemistry, hydrology, statistics, soils, etc. *Downstream research* on the other hand is a term used to designate applied and adaptive research, with watershed management *action research* as the far end of the continuum. Watershed managers are principally concerned with people's and society's problems. When watershed managers lack the information that is needed to adequately evaluate people's and society's alternatives, they often experiment with these alternative approaches until a satisfactory solution is obtained. Many watershed managers, therefore, become surrogate researchers in the process.

Next on the continuum are the watershed researchers who attempt to answer the more immediate questions asked by watershed managers – that is, those focusing on identifying resources capabilities, trends in the use of natural resources, as well as cost–benefit relationships and providing innovative technologies. Moving along the

---

[1] We are using this term broadly, as explained in earlier chapters, to include all the users of land and water in a watershed, including those formally appointed with the title watershed manager. They are people who make decisions on land and water use and are then responsible for carrying out those decisions.

continuum we come to developmental researchers who create new alternatives for watershed managers that, hopefully, can help to solve both immediate problems and those that will emerge in the future. Further along the continuum are the researchers who serve a clientele of other researchers. The former typically come from the basic disciplines to provide the facts, relationships and other inputs needed by those who conduct the more applied developmental research. Although the ultimate clients for all watershed researchers and managers are the general public, it can be seen that the intermediate clientele can be managers and/or other researchers.

Successful problem anticipation and, therefrom, effective planning become even more difficult as the distance on the spectrum between watershed researchers and watershed managers increases. While watershed researchers might focus on their immediate clients in many instances, their most basic activities are usually intended to help solve the broader management-oriented problems. It is important, therefore, that the planners of research anticipate these problems accurately and articulate their contents appropriately. Information necessary for the solutions of the array of problems confronted comes from managers, environmentalists and other members of society who are stakeholders in the enterprise. To the extent that this information comes from all of these sources, it confirms the premise put forth in this chapter that watershed research and management can – and perhaps should – be viewed as an iterative circular process of successive approximations.

Meeting people's needs

Watershed researchers and managers together face changing expectations on how the often-limited natural resources confronted should be used to meet people's needs. Management strategies, scientific knowledge and the technology necessary for producing the traditional multiple uses of natural resources only partially satisfied people's interests in the use of these natural resources through the 1990s. Watersheds were often characterized in terms of their capacities to yield demanded commodities and amenities. A primary role of watershed research at this time was discovering the factors that limited the realization of these capacities, while a key purpose of watershed management was reducing or removing these limitations. Answers to questions about the availability of natural resources also required the identification of optimum yields among the desired but frequently competing uses (Kessler *et al.*, 1992). However, this multiple use philosophy was not necessarily the most appropriate way to approach watershed management when people began to ask far-reaching questions on how to balance a wide range of potential uses with environmental quality.

In entering the 21st century, watershed researchers and managers have become increasingly more responsive to the more demanding viewpoints that people have of the natural resources on watershed landscapes and the respective contributions of these natural resources in meeting people's desires (Kessler *et al.*, 1992). Such a holistic perspective embraces a stewardship that balances a sustainability of products and services needed by people with protection of natural environments. It requires researchers and managers to view a watershed in a comprehensive context of a living system – including soils, water, plants, animals, minerals and all of the ecological processes that link these components together – that has importance beyond the traditional commodity and amenity uses. This viewpoint, often called the *ecosystem approach*, has been proposed by the National Research Council (1990) of the USA as a foundation of natural resources management into the future. We suggest that this ecosystem paradigm applied equally well as a focus on watershed research and management that meets people's needs.

A conceptual model that can be useful in establishing a research agenda that fills the necessary informational gaps to meet people's needs is presented in Annex 7.1.

Training Activities

The transfer of information and technology available through research endeavours and managerial experiences is integral to all watershed developmental activities. This is particularly important in the IWM training of all kinds since a successful IWM process is a participatory one that involves all the stakeholders, which means that all of them need some preparation and expansion of information and understanding, some of which comes from the various training activities. People possessing varying levels of knowledge and/or experience need information and, in many instances, skills and methodologies to plan, implement and then manage watershed management practices, projects or programmes (see Box 6.1 in Chapter 6 for a discussion of these levels of activities). Relevant and innovative information and technology relative to watershed management become available continuously, further requiring effective and efficient transfer mechanisms. Transfer of information and technology through training activities in a classroom setting and by distance learning to reach people (trainees) who are separated by location and time are two of the widely applied of these mechanisms.

Classroom training activities

Training activities in a classroom setting have been a standard approach to transferring information and technology and, it is likely, will continue to be essential in maintaining the necessary knowledge level for effective watershed management and operational functions. The primary function of this training is presenting specific information in relation to watershed management to trainees (students) and, in dong so, teaching functions and skills to enhance a person's ability to perform management, operational and/or research tasks related to IWM. Training activities involve exercises and practices that improve a persons skills, building upon the trainee's existing knowledge and encouraging the trainee to benefit from the experiences of others. Training entails concentrated activities designed to provide relevant information, key functions and appropriate skills to undertake IWM practices, projects or programmes.

Strictly speaking, training activities differ from educational programmes, with this difference reflected largely in the scope, depth and breath of the information and technology transfer (Box 7.1). Importantly, training and education are not mutually exclusive, but are rather oriented to satisfying differing levels of people's needs. However, training activities are the focus here.

Training levels

Because different groups of people need to obtain different types of information and technology, training in IWM is generally offered at either *strategic* or *tactical* levels

> **Box 7.1. Differences in training activities and educational programmes. (Brooks and Ffolliott, 1993, 2002.)**
>
> Training activities and educational programmes are both tools for transferring information and technology. However, they differ in the scope, depth and breadth of their transfers. Training teaches specific information, functions and skills to improve a person's ability to perform management, operational or research functions related to a specified discipline or objective. The focus of this chapter has been placed on training activities. Most training activities present basic concept while educational programmes can also employ a specific approach to reinforce basic concepts. In contrast, educational programmes focus on general principles and deal with all of the information available on a subject. Education is often associated with formal, broadly based, degree-granting programmes in watershed management, watershed hydrology and/or forest, rangeland or wildland hydrology.

depending on the specified needs of the trainees (Brooks and Ffolliott, 1993, 2002). Training at the strategic level might be aimed at sector planners or manager working at a national or regional level; senior administrators with a general awareness of the economic, environmental and social importance of IWM in relation to sustaining the flows of natural resources; or policy makers responsible for formulating and/or enforcing policies for guiding the planning and management of effective and efficient IWM practices, projects and programmes. Local leaders and administers of NGOs are often other groups to consider at this level of training. Training at the strategic level can also be directed towards convincing the interested public and society as a whole of the importance of supporting IWM programmes. Detailed instructions concerning technical methods and field practices of watershed management are not necessarily appropriate at this level.

Tactical level training at the *operational* level focuses on planners and managers and policy implementers to enhance their understanding of planning, appraising alternatives, managing and monitoring and evaluation. Tactical training might emphasize how to use watershed management practices to achieve both protection and production goals and can also consider institutional issues. This level of training is commonly directed largely toward field-level staff and technicians who need specific details on the methods of taking measurements, installing equipment and analysing data. Appropriate training topics might also include ways to implement soil conservation, construction of contour terraces and gully-control structures, and/or revegetation methods including species selection, planting techniques and protection. Training at the tactical level emphasizes knowledge gained, teaching methods and ways to adapt training materials to other situations. Training of trainers (instructors) is aimed at people who would use their training to help diffuse watershed management methodologies and practices to others, e.g. extension agents, can be the training goal.

A modular approach

Most of the training activities in watershed management consist of lectures, discussions, exercises, field trips and other learning methods, which we call *sessions*. These sessions can be grouped into *modules* according to their topical content and/or intended application. Modules, therefore, can be the primary components of a training activity. Experience has shown that a modular approach to watershed management training is both effective and efficient in terms of transferring information and technology (Brooks and Ffolliott, 1993, 2002). A trainer can use a modular approach in several ways to organize a training activity. For example, if a training activity is an overview of IWM practices, the trainer might opt to use a general lecture on this subject as the module. Additional sessions could then be added to this lecture as needed to devote more time to economic appraisals of IWM.

Depending on the purpose of the training activity, experiences and backgrounds of the trainees, as well as available time and resources, the trainer can formulate almost any combination of sessions from the array of basic modules. They can also put together a series of sessions in a building-block manner, designing modules to meet trainee needs. Many topics dealing with watershed management and its related planning, design and implementation have already been developed into training modules and sessions.

Illustrative modules and sessions for a variety of IWM training activities are presented in Annex 7.2.

Training formats

The format for training that satisfies a specific need depends largely on the training purpose and the targeted trainees. There are many training formats. A common format is a course – a format that many people commonly associate with training. A *course* is a series of training modules leading to a certificate of completion or its equivalent that relates to a particular well-defined IWM theme, topic or subject. A course can consist of lectures, examples, case studies, problem-solving exercises, field trips and other activities. However, there are other training formats that people might consider in planning for a specified training activity. Some of these other formats are discussed below.

The term *meeting* is a generic term referring to a gathering of trainees and, perhaps, other interested people for a designated purpose. Meetings can consist of a *conference*, *symposium* or *seminar*. While people often use these terms interchangeably, each has a specific meaning and a different purpose. A conference is a formal discussion or consultation on a range of subjects pertaining to a topic or general theme. A conference is a format for the interchange of viewpoints. A symposium is a conference in which a more specific and often narrower topic or theme is discussed and options are expressed concurrently. A symposium frequently consists of the presentation of a collection of opinions on a subject. A seminar is a meeting in which a group of trainees engage in discussions and exchange of viewpoints that focus on timely issues related to management, research or innovated ideas or concepts. It is conducted under the direction of a trainer. A *workshop* consists of a series of problem-solving exercises in which the trainees complete specific assignments in a classroom or laboratory setting. A *study tour* is a field trip or series of field trips that are conducted largely for demonstration purposes. It can be comprised of modules and sessions as a training activity or be used to complement other training activities.

A trainer (instructor) can select more than one format if necessary to meet a specified training purpose. For example, a conference on a selected topic that might be open to a broader audience of trainees could begin an IWM training activity. The conference could then be followed by a technical training course that is targeted to more select trainees, with the course consisting of a series of lectures on the theme and a workshop or study tour to provide the trainees with more specific skills relative to the selected topic. A trainer might also incorporate other training activities into the course as necessary to satisfy the overall purpose.

Learning methods

Sessions and modules can employ several basic learning methods grouped by subject into modules. IWM training activities typically consist of a combination of the following learning methods (Brooks and Ffolliott, 1993, 2002). *Lectures* involve the oral transfer of information from trainer to trainees that is often accompanied by view-guides, slides and/or other visual aids. Trainees with a limited background on the subject should gain new knowledge from lectures. *Lectures and discussions* also provide oral transfer of information to trainees, but facilitate closer interactions between the trainer and trainees and use the trainees' collective experiences to confirm information presented on the subject. *Panel discussions* are group interactions among trainees that are directed by a trainer and oriented toward a specific goal or purpose. *Small group discussions* are also group interactions but less formal than panel discussions. A trainer provides varying content and leadership aiming at problem solving, consensus building and decision making.

A *case study* is a report of a problem or situation by a designated person who has direct knowledge of the problem or situation. A case study is commonly structured as a detailed, in-depth discussion that is based on experience that considers all of the pertinent aspects of the problem or situation and attempts to point out the lessons learned. Parenthetically, a variety of case studies are presented in boxes throughout this book. A *problem-solving exercise* centres on the application of knowledge to a selected – sometimes hypothetical – situation to gain a practical acquaintance with a subject and/or skill with an activity. The exercises are often working sessions to practice applying the methods or knowledge that have been gained in the training activity. In a *structured role-playing* exercise, individuals within trainee groups assume assigned roles with different levels of direction and background information. Its purpose is directed toward making people more aware of how they and others might react in specific situations. A *game* or *simulation* involves extended, structured role-playing with specific rules and a larger trainee group. It helps to identify behaviours and attitudes in different situations. Physical actions that describe, illustrate or explain a subject by example is a *demonstration*. A *field trip* entails an excursion to a different location for demonstration purposes. It often complements other learning methods in the delivery of information.

Trainers use one or more of these learning methods to develop suitable sessions and modules for specific topics to be presented in a training activity. A key to identifying the most appropriate collection of modules for a particular situation is careful planning of the specific training activity (see below). Emphasis is often placed on local or regional conditions and viable approaches to achieve sustainable development through planned and properly implemented IWM practices, projects or programmes. The modules comprising a training activity would likely rely heavily on examples and case studies from the locale or region of the trainees in this instance.

Planning training activities

A successful training activity requires careful planning beforehand. The first step in this planning process is a *needs assessment* to identify who needs the training, that is, who the trainees will be; the purpose of the training to be conducted and the type of activity that best meets the training purpose. A needs assessment insures that the training fills a critical gap in the trainees' knowledge or skills. After identifying the trainees and otherwise completing the needs assessment, the sequential series of steps outlined in Fig. 7.1 are followed in the planning process. Importantly, the organizers of the training activity should also prepare contingency plans in the event that the successful completion of one or more of these steps is jeopardized in some way. A more detailed discussion on the process for planning of watershed management training activities is found in Annex 7.3.

It is helpful in planning a training activity of any format if the organizers are familiar with the learning stages in what is called the *adoption process*, and, hopefully, it is planned that the trainees pass through these stages in the training activity (Rogers and Shoemaker, 1971). A key in the adaption process is moving the trainees through the crossover threshold of this process – that is, from adaption and trial to adoption and implementation as illustrated in Fig. 7.2. Ideally, trainees would move from a position of relatively little knowledge of a subject to one of awareness, interest and understanding. The training activity would then build on the trainees' understanding of a

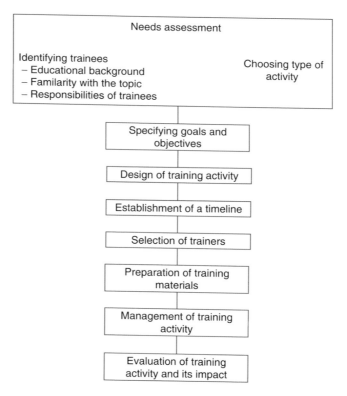

Fig. 7.1. Process of planning training activities. (Adapted from Brooks and Ffolliott, 1993.)

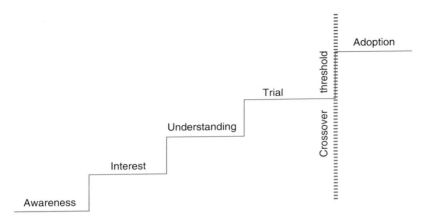

Fig. 7.2. Stages in the adoption process. (Adapted from Rogers and Shoemaker, 1971.)

problem and provide them with the technical knowledge and skills to formulate and carry out on-site solutions through trail and adaption if appropriate. The crossover threshold in learning represents the stage at which the trainees move from knowing how to do something to the point where the trainees will likely adopt and implement the training topics.

Because people generally learn in different ways – some by listening, some by seeing and some by discussion – the most effective and efficient method or combination of methods should be selected to move the trainees through the adoption process.

Evaluation of training effectiveness

It is always necessary to evaluate a training activity and its impacts on the trainees as thoroughly as possible in determining the effectiveness and lasting value of the activity. The scheduling of evaluations depends largely on the type of training activity selected and learning methods imposed; the duration (length of time) of the training activity; and how the evaluation process will satisfy the stated purpose (Brooks and Ffolliott, 1993, 2002). Evaluations of training activities might be scheduled at:

- Daily or weekly intervals to provide immediate insight into the training effectiveness and, when necessary, help in adjusting the content and methods to better fit the trainee needs;
- The midpoint of the training activity to determine if the trainee's needs are being satisfied, allow the trainees to re-evaluate their needs and obtain feedback to adjust content and methods for the remainder of the activity;
- The end of the training activity to provide an overall assessment of the content of the training activity to determine if the needs of the trainees, sponsors and other supporters of the activity have been met and, in many instances, help in designing future training activities on the subject.

Evaluations conducted 6–12 months after the completion of training are also useful in determining if the trainees still find the training received earlier to be relevant once they have been able to incorporate the newly acquired skills and knowledge into their work. This evaluation also establishes a 'mechanism' for staying in contact with the trainees.

Distance learning

Distance learning – also called *distributed learning* – is a way for a trainer (instructor) to reach a large number of trainees (students) that are separated by location and time. Both these terms have been used for decades in association with what are known as correspondence courses, where the trainees have relatively limited interaction with the trainer. Distance learning has become an important tool in improving the transfer of information and technology in relation to watershed management. It will likely become even more important in the future as computerized communicative technologies such as applications of the WWW technologies advance in sophistication and become more widely available to interested people (see below). Expanding the numbers and types of courses and developing new recruitment strategies are crucial and even though some strides have been made in that direction much more remains to be done.

There are advantages and disadvantages to distance learning that must be balanced against each other before implementing this form of technology transfer. Distance learning represents an avenue for reaching trainees unable to access standard training situations or as a supplement to that experience, although distance learning should not necessarily be viewed as a replacement for traditional institutions. One advantage of distance learning on the WWW is that this technology can provide the opportunity for trainees to preview a wide variety of courses and/or other training formats and choose the training that best fits their needs. Disadvantages to distance learning are that the trainee is separated from the trainer and other trainees; the interactive learning that takes place in a classroom environment can be missing; and a trainee does not necessarily have access to a hands-on laboratory. The use of interactive television (ITV) can partly eliminate the first disadvantage, although the trainer must be skillful in the use of this technology. Even if the training activity is offered entirely on the WWW, however, the classroom setting cannot be duplicated.

Recent innovations

Distance learning technologies are continuously advancing to improve the quality of the total learning experience and opening the doors to trainees previously excluded. The technological media available today include voice-activated ITV, long-distance conferencing and a variety of other WWW applications. ITV – also called video-conferencing – allows for easy interaction between the trainer at the originating site and a trainee at a remote site. Current systems can facilitate two-way video and audio communication so that the trainer and trainees can see each other and speak to each other as though they were in the same classroom setting. Electronic mail (e-mail) can also be used for interaction between the trainer and trainee outside the classroom for the submission of assignments, questions and ongoing discussions (see section on Internet Applications).

The WWW allows the trainer to post tests, examinations and visuals for trainee reading, study and response. These types of technologies also have the advantage of being compatible with the daily work of the trainees. Also possible through this medium are feedback mechanisms that allow self-testing, direct linkages to related instructional materials and instant access to glossaries and reference sources. In the future, it is likely that the WWW will provide even more personalized versions of trainer–trainee interactions, for example, two-way real-time video imaging – which is currently prohibitively expensive in many instances – is a means for face-to-face contact between the users.

Use of compact discs

Another approach to distance learning is through the use of compact discs (CDs) in offering a training activity. This technology offers more flexibility and the possibility of reaching a wider audience of trainees than those discussed above. There are areas in the USA and elsewhere in the world where people do not have ready access to the Internet. This situation can prohibit or restrict people's opportunities to further their professional training without leaving their workstations for a classroom. In many cases, these limitations or restrictions are financial with the non-availability of funds constraining a linkage to the Internet. However, within these limitations, a course or other form of training can be offered almost anywhere and at any time by utilizing CDs. In reference to offering formal courses on topics related to IWM, it is apparent that the use of CDs would allow for these courses to be tailored not only for the individual trainee but also for the area where the trainee resides (Box 7.2).

Internet Applications

Making IWM technology and other information available to a wider audience of people can often be enhanced through the use of the Internet – often as a supplementary

---

Box 7.2. Distance learning: a watershed management course.

A distance learning course in IWM is offered by the University of British Columbia, Vancouver, Canada, to practicing professionals, college-level students and other people whose work and/or interest involve natural resources management and who need the convenience of distance study to increase the level of their knowledge. Sessions and modules comprising the course include methods, techniques and tools; hydrology, sedimentation processes and water quality and impacts on aquatic biota; land-use issues in a watershed management context; a community-based approach to implementing watershed management activities; and governance and institutional frameworks confronted by people interested in watershed management. A listing of 400 technical references searchable by authors and keywords is provided to supplement the structured coursework offerings. The course has been organized in an interdisciplinary format, with a number of case studies incorporated into the offering to reinforce specific sessions. The case studies presented are related to issues of sustaining high-quality surface water and groundwater resources while controlling (mitigating) pollution problems caused by forestry, agricultural and urban activities. While the course is largely CD-based in its offering, the trainers (instructors) also use e-mail communications and an Internet bulletin board for 'interactive discussions' with the trainees (students). More details on this course are found on http://www.cstudies.ubc.ca/dipcert/watersh1.htm.

---

tool coupled with other learning methods – if the trainees have ready access to this technology. Appropriate applications of the Internet complements face-to-face training when planned properly. The Internet can also be a focus of distance learning formats when it is not feasible to bring people to a training site. It has become a useful tool that is increasingly helpful in the transfer of IWM technology in either case. A primary use of the Internet in transferring information and technology on the practice of IWM assists people (the users) by reducing the complexity that is perceived to be associated with understanding and implementing IWM (Figallo, 1998; Lawrence and Giles, 1998; Huebner *et al.*, 2000; Johnson, A. 2000). The Internet is a tool that allows communication among professionals and practitioners – of importance to the readers of this book – and the general public. It has proven to be an aid in removing barriers to understanding IWM by providing an online way to visually display information in a more comprehensible manner.

The Internet itself does not contain technical information related to IWM or any other topic of interest to people. It is a misstatement, therefore, for someone to say that information is found on the Internet. Rather, to say that the information obtained is found *through* or *using* the Internet is a more correct statement. More specifically, the information is found on (or in) one of the computer systems that is linked to the Internet. Internet communications technologies that are frequently used by watershed managers and the public to exchange information and technology include e-mail, bulletin board communications, blogs (weblogs) and the WWW.

Email

The most widespread form of Internet communication is e-mail. E-mail messages are a useful and easy-to-use tool for delivering technical content and receiving feedback of watershed-related information. E-mail messages are generally brief in their content and focused on a specific subject matter. A form of e-mail that is often used by watershed managers to send messages of common interest to an assemblage of receivers is a subscriber mail list, which is a collection of e-mail addresses of people with a specified interest in a topic. When the volume of e-mail messages becomes unwieldy, however, bulletin boards and/or blogs are often the mode of communication on the Internet.

Bulletin boards

Bulletin boards are established so that a conversation on a specified IWM topic is maintained and people (trainees) in the communication exchange can go online at anytime to read, study and respond to discussion comments to achieve this purpose. Bulletin boards, also called message boards, conferencing systems and/or asynchronous discussion forums, are effective when archived conversations need to be accessed for future reference and training consideration. In contrast, so-called *chat conversations* take place in real time and the conversations are not necessarily archived for later reference.

Blogs

Another Internet tool that has appeared in the last 10 years is a *blog* – the shortened name for a weblog. A blog is a specific type of web site where entries are made and displayed in reverse chronological order, that is, the most recent entry is displayed first, the second entry is displayed second, etc. Although blogs are sometimes a personal diary or journal relating to the activities of an IWM activity, they are increasingly used to provide new, commentary and/or other information on a particular subject of interest. Blogs are another asynchronous discussion forum that can be used, for example, to prepare (create) collaborative proposals or reports concerned with IWM topics.

World Wide Web

The WWW incorporates all of the Internet services mentioned above and more. A user can retrieve publications and other documents; data sets, images (photographs, line-drawings, etc.), software programs; and comprehensive bibliographies related to an IWM topic. This information is displayed on many web site formats (Box 7.3) that operate on almost any software in the world, if the user's computer has the

---

Box 7.3. Managing arid and semi-arid watersheds: one example of a web site format.

Information from the watershed management research and operational activities in north-central Arizona has been incorporated into an easily accessed web site (http://ag.arizona.edu/OALS/watershed/index.html) entitled 'Managing Arid and Semi-Arid Watershed' to provide a unique reference and a tool for disseminating relevant information on the effects of natural and human-induced disturbances on the functioning, processes and components of the ecosystems found in the arid and semi-arid environments of north-central Arizona (Huebner *et al.*, 2000; Haseltine *et al.*, 2002). Among the data sets on this web site are:

- Precipitation, air temperature and relative humidity regimes;
- Streamflow amounts, peak flows and timing of flows;
- Sediment yields and, more generally, water quality characteristics;
- Timber growth and yield and forage production;
- Wildlife populations and preferred habitat qualities.

There are links to lists that allow the user to access the diversity of categories contained on the web site. *Search lists* of the types of information contained on the site are available to the user. *Drop-down lists* are also available for easy access to data sets for selected watershed conditions and specified years. An *overview* provides a narrative on the Beaver Creek watershed project, the focus of numerous IWM research effort in north-central Arizona (Baker, 1999). Included in the narrative are a site description and history and highlights of the research findings. *Data* categories include precipitation and other weather features; soil characteristics; streamflow regimes, sediment concentrations and water quality constituents; timber and forage production, and wildlife population estimates and habitat conditions. An *image* file contains nearly 2500 photographs and other illustrations on a variety of watershed-related topics. A *publications* link contains nearly 700 annotated citations of publications and reports on the Beaver Creek project. The information and technology on the Beaver Creek web site provides a basis to help watershed researchers, planners and managers, as well as policy makers resolve current and future land stewardship issues.

---

hardware and software to do these things. Web sites are typically controlled by either a formal organization or a group of people. Therefore, while users can access information from the web site, they (the users) cannot directly contribute information to the site.

Subject guides and *search engines* are usually available to help watershed managers and the public to effectively gather and distribute this information. Subject guides are hierarchically organized indices of subject-matter categories that allow the user to browse through lists of available web sites by the subject matter of interest in search of information. Some subject guides are general while others are specialized. General subject guides are suitable for exploring informational resources about broad topics such as arts and humanities. If the desired information is more specific in nature – such as the streamflow records for a watershed located in a specified biogeographic region – specialized search engines can be helpful. Because there are literally thousands of specialized subject guides available, *clearing-house sites* are often available to help the user in efficiently accessing them.

Search engines are developed by building an index from existing web sites and then providing the user with the ability to query that index. To build a desired index, search engines deploy software robots that automatically index the contents of a web site. A robot indexes the WWW pages that are linked to the first page and then moves on through cascading myriads of linked pages. Because of the automation used in their development, search engines can index a larger portion of the WWW than subject guides. A larger index means more pages relating to a narrowly focused topic are found and delivered to the watershed manager. However, because search engines index many web sites, a large portion of the pages can lack relevancy to a selected topic – this is especially true if the query is overly broad.

There are a number of ways by which a person can find information on the WWW. One approach is taking advantage of subject guides and search engines. Another way of obtaining needed information is attaching the address of web sites to communications with colleagues. Information about WWW pages on watershed management topics is also found in newsletters, bulletin boards and mailing lists. Web sites have become the universe of network-accessible information and technology for IWM.

One word of caution on the use of the WWW is warranted. Users of the technology need to keep in mind that there is no quality control for the information on the WWW. Almost anyone can purchase space on a WWW server and post a document on almost any topic. Because a document is found on the WWW does not necessarily mean that the information contained in the document is factual or correct. It is, therefore, important for users to pay particular attention to the source of the information found on the WWW, this is, whether the web site is a personal site or the document of interest is posted through an educational institution or a government organization.

8 Adaptive, Integrated Management of Watersheds: Concluding Thoughts

Due to the many internal and external linkages between actions taken in a given watershed and because many of the actions taken in a watershed produce both positive and negative impacts that go far beyond the boundaries of that watershed, people need to think broadly about the challenges confronted and opportunities that exist on a watershed or river basin (see Chapter 1). Such broad thinking should lead to actions that take advantage of the opportunities presented for positive interactions and control the negative ones. This means thinking nationally, regionally or globally in some cases, and concentrating at the watershed or river basin level in other instances. Most environmental issues that have national, regional and global implications are ultimately resolved only by dealing with the issues at the local level as defined within the framework of the existing physical and institutional boundaries. While a national government can make a decision on a solution to problems such as large-scale soil degradation, water quality improvement or deforestation, we know that these solutions only work if people act locally.

Local actions can take place at various levels – farm fields, urban development sites, forest lands, grazing lands and so forth. Because of the biophysical and socio-economic interrelationships within the watershed or the river basin boundaries, people must recognize that their actions can impact others – and that there can be many stakeholders involved. Institutions must exist that help resolve conflicts over land and water. Achieving equitable, efficient and effective resolution of issues is facilitated if there is proper accounting of positive and negative cumulative effects associated with a changing landscape. Local actions are needed, but planning and actions on a broader basis are also necessary, for example, at the entire watershed or river basin level, or even at the national global levels. Only by considering the small watershed within its larger context is it possible to identify and deal with the many interactions that take place within and beyond the watershed boundaries. Matters can become quite complex as the scope of planning and actions expand out. Thus, all of the needed planning and actions should be based on the best analysis possible of what has happened in the past and the lessons, both positive and negative, that can be learned from past management results. Such lessons, and the ways in which they can be utilized in planning and implementing future management are at the core of a process known as *adaptive management*.

Role of Adaptive Management

Although people might plan for specific natural resource outcomes and specific environmental and welfare effects when IWM is being implemented, they cannot always be sure that the envisioned activity will be carried out as planned nor that the activity will have the effects on the people as anticipated. This dilemma arises largely because of uncertainty regarding the effects of land use and watershed management practices on land, water and other natural resources; climatic conditions; changing demographics; and the general welfare of people involved. An *adaptive management approach* provides a means to cope with these uncertainties. Adaptive management involves systematic monitoring and evaluation of past experience and then introducing the lessons from such evaluation into adjusted goals, plans and implementation. Adaptive management helps to guide – when it becomes necessary – changes in land use, watershed management practices and relevant planning processes and policies that are necessary to assure attainment of the desired goals of watershed management. Monitoring and evaluation are integral to provide the feedback necessary to make adjustments in planning and management when necessary. Thus, adaptive management comes into play throughout the processes of planning, monitoring and evaluation, as well as implementation of watershed management practices on the landscape.

Adaptive management provides a good general strategy for connecting more effectively people to their land and water over time. The underlying theory and concept of adaptive management – people learning from experience and then modifying their behaviour in light of that experience – is not new. However, it has been formalized and has gained increasing attention and interest in the context of its potential application in the planning and implementation for the management of land, water and other natural resources in recent years (Holling, 1978; Walters, 1986; Lee, 1993; Peterman and Peters, 1998; Murray and Marmorek, 2003; Stankey *et al.*, 2005). The adaptive management approach is consistent with, and in fact is part of what we have called IWM throughout this book. The foundational premise, components and framework and alternative models for applying the adaptive management approach are discussed in this concluding chapter.

Foundational Premise

Watershed management practices such as have been discussed in this book contribute to sustainable flows of natural resources and involve specific activities that use, affect and are affected by the environment (see Chapter 2). The health of the environment (in turn) affects economic and social activities and results and, ultimately, the well-being of people. The purpose of planning and implementing IWM activities within a holistic framework is to change land-use practices and associated activities to improve the welfare of present and future generations of people without affecting the environment adversely. However, the linkages among watershed management, environmental systems and the welfare of people are impacted by climatic, social and other changes that cannot often be predicted.

In paraphrasing the discussion of Walters (1986), Walters and Holling (1990) and Stankey *et al.* (2005) on adaptive management in general, a foundational premise for proposing adaptive management of land, water and other natural resources is that technical, social and economic knowledge of the watershed or river basin in question is usually incomplete and often elusive. Furthermore, expanding the necessary knowledge of these topics through scientific inquiry can be limited by resources and time. When these limiting factors are linked to conditions of natural resource scarcity, possible irreversibility and increasing demands for benefits, the need for a better way of understanding the decision-making and policy processes involved (see Chapter 4) becomes apparent (Bormann *et al.*, 1994, 1999; Stankey *et al.*, 2005). Adaptive management offers a scientifically sound course of action that does not necessarily require that action is largely dependent on extensive study and a strategy of implementation designed to enhance systematic evaluations of the actions involved.

Components of the Process

Adaptive management acknowledges that resources and time are often too short to defer some actions. At the same time, it is also true that actions are often postponed until enough is known about past experience to actually implement the actions with some degree of knowledge regarding likely outcomes and impacts. The major components of the adaptive management process (Holling, 1978; Walters, 1986, Stankey *et al.*, 2005) are:

1. Integration of existing knowledge and experience into dynamic models to frame predictions about the impacts of alternative policies and actions – this component performs three functions:

- Problem clarification and enhanced communication among researchers, managers and other stakeholders;
- Policy screening to eliminate options unlikely to achieve success because of inadequate type and scale of impacts;
- Identification of key knowledge gaps that make predictions suspect.

2. Design of alternatives for a specific management practice.
3. Linking of the results of a management practice with the policy-making process – that is, determining how these results translate into changes in ongoing watershed management practices in light of the actions taken.

Implementing changes in land and water use aimed at meeting the objectives of IWM can bring about an array of possible outcomes no matter how carefully planned and implemented. For example, circumstances can change dramatically from the time planning begins until the changes are implemented (see Chapter 4). Therefore, we cannot be certain that:

- A proposed watershed management practice will be implemented as planned;
- The watershed management practice will have the anticipated effects on the land, water and other natural resources and human activities;
- The natural resources, environmental and other changes that are brought about by the watershed management practice will affect people's welfare as originally anticipated.

Adaptive management has often been suggested as a logical approach for 'watershed and water resource planning, management, and restoration…' according to Stankey *et al.* (2005), Sehlke (2006) and St. Clair *et al.* (2006). In essence, adaptive management represents a number of alternative approaches for managing effectively and efficiently complex environmental systems such as watershed landscapes and coping with the inevitable changing circumstances encountered as people use the resources on the watershed.

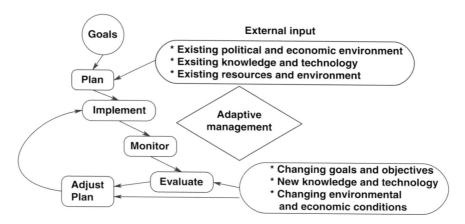

Fig. 8.1. Adaptive management cycle. (Adapted from US Department of Agriculture: US Department of the Interior 1994, Stankey *et al.*, 2005.)

Framework of the Process

The steps that form the framework for the adaptive management process are illustrated in Fig. 8.1 by a four-phase cycle involving the four basic elements: planning, implementing, monitoring and evaluating. A plan is framed in the first phase of this cycle. This plan is based largely on the organizational goals to be met, available resources, current knowledge of natural resources capabilities and suitabilities, existing and feasible technologies relevant to management and available information from inventories and monitoring and evaluation summaries. On-the-ground actions – in this case an IWM management practice – are implemented in the second phase of the cycle. Phase three involves monitoring the ongoing processes and the results (outcomes) of the implementation. Phase four is an evaluation of the effort which then leads to possible adjustments in the initial plan based on changing goals, and changing external conditions related to the environment, new technologies, resource availability, etc. This four-phase cycle is then repeated with revised goals and plans, if in fact such revisions were found desirable and feasible. It is what we would call an iterative process of successive approximations as new information and technology, changing goals, changing conditions and changing resources come to light. It is a dynamic process, just as the Brundtland Commission defined *sustainable development* as a dynamic process rather than an end state to be achieved. The final results of the adaptive management process can validate existing ongoing practices and policies or reveal a need for alternative policy management adjustments.

As indicated in previous chapters, M&E are essential throughout the adaptive management process to provide the necessary and updated information for evaluating and making adjustments. In large part, M&E is an iterative process that allows people (stakeholders) to identify the necessary changes in an IWM practice to achieve the goals and objectives that were intended to be achieved in the first place (see Chapter 6). The adaptive management process further supports the notion that because a specific plan can evolve through time, this should be viewed as a working document and not necessarily as a prescription that must be blindly followed in its implementation. As mentioned earlier in Chapter 4, the use of a 2- or 3-year moving plan makes a lot of sense if conditions are changing fairly quickly. Such an adaptive management plan involves, for example, developing a plan for 2007–2009, with the next, incrementally adjusted plan being prepared the following year for 2008–2010 and so forth. Each year (e.g. beginning of 2008 in the case of the 2007–2009 plan) one can: (i) look back on the results of the first year, e.g. 2007 – checking the achievement of benchmarks and objectives that were established a year earlier; (ii) be more certain about the coming year (2008) and make necessary adjustments and more detailed plans and benchmarks; and (iii) adjust the plans for the third year (2009) accordingly and on the basis of additional information and changes in resources that have occurred since the last plan. These adjustments then lead to the establishment of the new 2008–2010 plan. This type of medium term planning (as we called it in Chapter 4) is in common use in some organizations, e.g. in the 15 major international agricultural research centers that make up the Consultative Group on International Agricultural Research.[1]

Alternative Models

Walters and Holling (1990, as reported by Stankey *et al.*, 2005) suggest that there are three models in which adaptive management can be structured and achieved. There is an evolutionary or trial-and-error model in which the results of external decisions and choices are used as a basis to structure subsequent decisions that – hopefully – lead to improved results. This model of adaptive management is reminiscent of the learning that results from whatever management experience is undertaken (Stankey *et al.*, 2005). In some quarters, it is called an *institutional learning and change* (ILAC) model. There is little or no purposeful direction to the use of this model and – to a large extent – people simply reap the benefits that are derived from earlier experience.

Passive adaptive management is a second model. Bormann *et al.* (1999) used the term *sequential learning* to describe the form of this model in which historical knowledge and information are used to frame a single and best approach to management along a linear path that is assumed to be correct. This model applies a formal and rigorous – albeit post facto – analysis to secondary information and experiences as a means of structuring new choices, understanding and decisions. However, there are two problems that can limit the usefulness of the passive approach to adaptive management (Stankey *et al.*, 2005). First, this approach can often confound management and environmental effects (outcomes), because it is unclear whether the observed changes following the implementation of a management practice are the result of how the land, water and other natural resources were treated or caused by changes in the environment such as the possibility of global warming. Second, this approach can fail to detect adequately the opportunities for improving the system performance when both the right and the wrong model predict similar results and the system is managed as though the wrong model is correct.

The active adaptive model is the third model to structure and achieve the adaptive management approach. It differs from the other models in its purposeful integration of experimentation into policy and management design and implementation

---

[1] Cf. http://www.sciencecouncil.cgiar.org/publications/pdf/mtpguide06.pdf

(Kusel *et al.*, 1996, as cited by Stankey *et al.*, 2005) – that is, policies and management activities are considered to be experiments and opportunities for learning (Lee, 1993). Active adaptive management is designed to provide information and possible feedback on the relative efficacy of alternative policies and management activities rather than focusing the effort on a search for the single best predictor (Stankey *et al.*, 2005). Bormann *et al.* (1999) have referred to active approaches to adaptive management as examples of parallel learning because they involve the design of a series of policies that can be compared and evaluated. *Action research* is often associated with this model, i.e. applying the scientific method as actual management and change is taking place and using the IWM framework as a laboratory for learning.

The literature presents a variety of ways to undertake the adaptive management approach but with no templates to guide people about the best model (Stankey *et al.*, 2005). Nevertheless, the central elements that the adaptive management approach seeks to foster include formal learning coupled with forums that facilitate improvements in problem identification and framing; mutual and ongoing learning, and an informed debate about the alternatives offered.

Applying Adaptive Management

The presumption in applying adaptive integrated management to land, water and other natural resources is that people have the necessary resources, ability and authority to make the necessary and timely changes as IWM practices become implemented and people begin to observe results. Even when an adaptive management response seems to be clear, the complexity of watershed ecosystems and their response to alterations of land, water and other natural resources, possible climatic variability, and social and economic responses can require further – and perhaps continuing – adjustments over time. As Thomas (2006) indicates 'land management plans, promulgated under the adaptive management concept, are simply hypotheses in which proposed management actions are expected to produce results with the anticipated ecological impacts'. Through implementation, monitoring and evaluation, people are able to test these hypotheses.

The adaptive management approach applies to large-scale programmes such as the USDA Forest Service National Forest Planning effort that began in the early 1970s, programmes on river basins and smaller watersheds, and even activities at the individual user level. Although described in theory and concept by many people (references listed earlier), the idea embodies a common sense approach – that is, plan it, do it, and if the desired results are not forthcoming or unwanted results are observed, change the course of action as dictated by the circumstances. What is unique about applying adaptive management in an IWM framework is that the results or impacts can be analysed both on-site and off-site or downstream and can, therefore, be readily observed and/or recognized by those people involved with land-use change or watershed management activities.

Conclusions

Some variation of adaptive management provides the means of more effectively connecting in a productive and sustainable fashion people to their land and water, utilizing the types of information provided in the previous seven chapters. At the same time, we must also recognize that there can be barriers to the use of such a flexible and dynamic management model. This is particularly true in larger organizations – particularly government entities – that often have to live by bureaucratic rules, deadlines, prescribed management methods and planning horizons – ones that often follow the election cycle. Nevertheless, as IWM gets more widely adopted and put into practice by local quasi-governmental and civil society groups, where participation, cooperation, collaboration, flexibility and adaptability are the cornerstones, adaptive management can be a potentially important – even essential – part of the efforts of people to deal more effectively with today's complex and often uncertain world of connecting people in a productive and sustainable way to their land, water and other natural resources (Holling, 1978; Walters, 1986; Lee, 1993; Stankey *et al.*, 2005). Without some form of adaptive management approach, rule-based planning will likely continue to dominate the management of land, water and other natural resources in the future, with a continuing diminution of the ability of responsible people and interested stakeholders to modify actions and policies in light of new knowledge and experience. Short-term economic interests could very well dominate at the expense of the environment on which longer term, sustainable economic activity depends.

Moving ahead in the right direction will require expanded commitment, improved partnerships, innovative financing, continual technical innovation and leadership from land owners and managers, policy makers, scientists and the general public. An adaptive, IWM framework will also be needed to ensure an expanded, productive and sustainable use of land and water. The phenomenal growth in some countries of local and regional watershed and river basin councils, associations and other entities that have heavy influence from civil society, land owners, businesses, as well as government agencies is extremely encouraging, since many of them embody the principles that are necessary to ensure sustainable, integrated management of watersheds and river basins. They also provide evidence that is becoming clear to most of us that we all are stakeholders in the watersheds on which we live and that all of us, whether landowners or not, have an influence on the availability of quality water in quantities sufficient to satisfy our various needs. Moreover, all of us have a role to play in ensuring continued productive use and enjoyment of our watersheds and the water that flows through them and lies under them. The principles for success have become quite clear and, as indicated in Chapters 3 and 4, although they are voiced differently by the many varying groups drawing lessons from their own experiences, they basically all boil down to some logical, albeit obvious requirements for good adaptive IWM as follows:

- Widespread commitment and participation from people in the watershed or river basin.
- Partnering, consensus building and sharing of decisions and responsibilities between government and civil society entities, geared toward strengthening leadership qualities and giving clear consensus direction, both of which are necessary for progress.
- Expanded information sharing, education and training, taking maximum advantage of the new ICTs, as discussed in Chapter 7.
- Clear recognition and understanding that technical and institutional strengths are like the two blades of a pair of scissors – one without the other does a poor job of getting the cutting done properly, as stressed throughout the book.

- Creativeness in developing innovative financing mechanisms and volunteerism – sharing of expenses and actions – following the polluter pays, and user pays principles where possible.
- Flexibility and adaptability in planning and management – learning by doing; use of flexible adaptive management approaches.
- Expansion of incentives, innovative systems of payments for environmental services that are outside the market place, and more responsibility to the private sector to follow good corporate codes of conduct, coupled with a significant reduction of government command and control approaches.

These lessons, principles, guidelines or whatever one wants to call them are easy to recite, but often difficult to implement in the imperfect world in which we all live. Yet, it is encouraging to see that people are waking up to the possibilities for sustainable management of the resources on which we all depend for livelihood and welfare. The growth of local interest in and concern for the watersheds on which we all live is an encouraging sign. There are many thousands of formally organized local watershed management councils, associations and other action groups. Yet, despite this encouraging sign, with the majority of us living in cities and far removed and isolated from the sources of what we use every day – food, fuel, wood and other raw material-based housing, furniture, water and so forth, we often do not know, or at least forget that ultimately availability of all these products depend directly on how we manage and protect the natural resources on and under the watersheds that determine the quality and flow patterns of the most important natural resource needed for our survival – water.

Annexes

Annex 3.1

A Process to Identify, Assess and Deal with Policy Issues

Human-caused land and water degradation is accelerating in many parts of the world and, as a result, a decline is occurring in the production of food, fuel, fodder, fibre and safe water needed for the basic welfare of people living in these regions. This degradation of the environment – whether through deforestation, desertification, soil erosion, groundwater depletion and/or pollution – is creating some of the more pervasive policy issues facing people today (Gregersen *et al.*, 1994b; Quinn *et al.*, 1995). By improving the policy environment, people are often able to contain or reverse these declining trends and move themselves toward a more stable and sustainable level of living. However, governments and development agencies are faced with difficult choices in dealing with these issues, because they usually involve many sectors of society, government agencies and stakeholder groups. It is crucial, therefore, that policy makers apply systematic and transparent processes to assess these issues and develop policy options to resolve them. Such a process by which key natural resources policy issues can be clearly identified and assessed and feasible options can be developed to deal with these issues is presented in this annex.

Natural Resources Policy Issues

Four key points must be kept in mind about policy issues related to natural resources and their use. They are:

- Policy issues involve different perceptions by different people (stakeholders) of the best uses for natural resources. Different uses will have different values and different degrees of acceptability for different people.
- Policy issues, therefore, involve disagreements and/or controversy among competing interests over the current and future uses of natural resources and the associated goals and objectives to be satisfied by such uses.
- Policy issues are different from managerial or technical problems in terms of how they should be addressed. In the case of a management issues, experts can analyse and make recommendations for resolution of the issue using criteria established to meet the goals and objectives specified in the existing policy. In the case of policy issues, it is disagreement over these goals and objectives that leads to the emergence of the issue. Different value frameworks exist (Gregersen *et al.*, 1994a). Therefore, different sets of criteria for choice among options to resolve an issue also exist. These are differences that need to be addressed.

- Policy issues are dynamic and evolve over time as views on the uses of natural resources change. For example, land conservation was not much of a policy issue in the early 1900s in most countries. However, it became a major issue in later years as incomes grew, technology advanced, relative political power shifted and changes occurred in the values and influence of environmental groups.

The political process and the role of government

Patterns of natural resources generally use change in an incremental fashion over time through power shifts and negotiation and compromise among users of the natural resources, governments and other interested stakeholders. It is essential that policy makers act as advocates for the people who have no means to represent themselves in a policy debate in this process (Gregersen *et al.*, 1994a). These people are generally the disenfranchised poor. Resolution of policy issues in this context comes through political advocacy, bargaining and trade-off processes. Implications of alternative value systems are debated, negotiated and forged into compromises that are most acceptable to society at the time. The most that government analysts and/or planners can do in most instances in the debates and negotiations is helping to clarify the perceptions of the involved people regarding values and, therefrom, the issue. However, these analysts and/or planners can also provide an assessment of the policies that are related to emergence of the issue; suggest options for changing these policies; and furnish insights on society's and government's capacity to implement each of the options developed.

There is absolutely no best solution or right answer in a debate over policy issues in a democratic advocacy society. Rather, there are possible compromises between what different stakeholders in the issue view as right or wrong, or acceptable or unacceptable. A good solution is one that equitably addresses the interests of all stakeholders – including those without the ability to effectively represent their own interests – and reduces any conflict below the threshold of concern to policy makers upon its implementation. The process that is presented in this annex can be used by policy makers and their advisors to assess the issue in a way that can help them forage such compromises (Gregersen *et al.*, 1994b). It does not lead to definitive answers to questions such as 'what is best?', 'what is right?' and 'what is just?'

An overview of the political process of resolving natural resources use issues in an advocacy society is discussed in Chapter 4. In that chapter, Fig. 4.1 provides an overview and the political process is shown in the left column. A parallel policy assessment and policy design process is illustrated in the right column of Fig. 4.1. Government agencies are largely responsible for this process to support decision making. Note that it is an interactive and iterative process as successive stages in negotiations are reached. The support functions of government analysts and (or) planners include the elements of concern presented in this annex.

Issue Assessment and Policy Design

Some emerging issues cannot always be resolved within the existing policy framework. These issues emerge as policy concerns, with agencies often entering into the stage of policy assessment and policy design (Gregersen *et al.*, 1994b). Sequential stages in this process are:

- *Issue assessment: defining the issue, its impacts and its causes.* This first task is identifying and defining the exact nature of the issue and the activities that lead to its emergence. This task is a critical step since resolution depends on the correct identification and definition of the issue. It involves an analysis of why the issue is an issue and how it got on the agenda. At this stage, one also needs to identify and clearly analyse the value perspectives of the different people and/or groups involved in the issue. Questions that might be asked are, 'from whose point of view are the issues a problem?', 'who benefits and who gains?' and 'what are the opportunities and threats that the issue poses?' To fully understand the issue and how to resolve it, we need to identify the underlying reasons why the controversial activities occurred that lead to its existence.

- *Policy assessment: assessing policies that are associated with the issue.* The purpose of this stage is to fully assess the existing policy context and its impacts on stakeholders and their actions before designing new (alternative) options. The focus at this stage is placed on identifying the policies that created the environment or context in which some of the stakeholders engaged in the controversial activities and some of the stakeholders vocally and formally oppose the activities. Two types of weaknesses in policies are considered in this regard. First of all, there are ineffective policies – that is, the intention of the policy might be to accomplish a specified policy objective but it ends up not accomplishing that objective. For example, a policy might have the objective of reducing the pressures on harvesting of natural forests by encouraging tree planting through subsides and/or technical assistance. Instead, the policy results in the establishment of tree plantations in areas where there is no demand for wood from natural forests and, as a consequence, pressures on the natural forests continue as ever. Second, there are conflicting policies – that is, where a policy in one sector has unintended and unanticipated negative effects in another sector or a policy intended to have a specified positive impact also has a negative impact. For example, a policy aimed at providing additional lands for poor people results in the loss of critical biological diversity through deforestation activities by the people. Or, a policy aimed at promoting exports through the provision of price supports for agricultural exports results in the deforestation of lands that otherwise would have been marginal for agriculture but critical for watershed protection. Considering what we have learned about the policies associated with the issue of concern, we next identify – at least in a preliminary way – the key policies that require policy intervention. This task is a transition to stage three.

- *Policy design: assessing and developing options for trade-offs and compromises.* The purpose of this stage is to define and assess the alternative options available for dealing with the key trade-offs surrounding the issue. We attempt to identify the potential options in each problem area and how these options might influence the attitudes and actions of the competing people or groups of people. We would also examine the sustainability, equity and economic efficiency dimensions of each option. Questions to be answered include, 'How sustainable are the implied changes?' 'Who would be affected by the changes and how?' 'What would be the costs of each option in relation to the perceived benefits from society's collective point of view?' and 'How – if at all – should society compensate the people who "lose" in each option?'

- *Organizational assessment: assessing implementation needs, capabilities and feasibility.* The purpose of this final stage is assessing the likely feasibility of implementing the main options. This stage involves considering the following questions: 'What are the key organizations on which the success of a policy option depends?' 'What is expected from them?' 'Can the key organizations implement the policy?' 'In what ways do key organizations need to change?' and 'What are the resource, knowledge and incentive constraints that need to be addressed?' Answers to these questions tend to be unique to each institutional environment. Organizational feasibility is highly dependent on the specific circumstances surrounding the issue and the nature of the institutional environment in which it exists.

Conclusions

The issue assessment and policy design process outlined in this annex illustrates a process for addressing issues in a systematic fashion. This assessment process is neither new nor original and that is its strength. It is a time-tested and accepted process that provides information for applications in the more complex and unpredictable political process of issues resolution.

Bibliography

Gregersen, H., Arnold, J.E.M., Lundgren, A. and Contreras A. (1994a) *Valuing Forests: Context, Issues, and Guidelines.* EPAT/MUCIA/USAID Draft Policy Brief, University of Minnesota, St. Paul, Minnesota.

Gregersen, H., Brooks, K., Ffolliott, P., Lundgren, A., Belcher, B., Eckman, K., Quinn, R., Ward, D., White, T., Josiah, S., Xu, Z. and Robinson, D. (1994b) *Assessing Natural Resources Policy Issues.* EPAT/MUCIA/USAID Draft Policy Brief, University of Minnesota, St. Paul, Minnesota.

Quinn, R.M., Brooks, K.N., Ffolliott, P.F., Gregersen, H.M. and Lundgren, A.L. (1995) *Reducing Resource Degradation: Designing Policy for Effective Watershed Management.* EPAT/MUCIA/USAID, Working Paper 22, Washington, DC.

Annex 3.2

Principles and Standards for Privatization
(Excerpts from Gleick *et al.*, 2002)

We believe that the responsibility for providing water and water services should still rest with local communities and governments, and that efforts should be made to strengthen the ability of governments to meet water needs. As described in this study, the potential advantages of privatization are often greatest where governments have been weakest and failed to meet basic water needs. Where strong governments are able to provide water services effectively and equitably, the attractions of privatization decrease substantially. Unfortunately, the worst risks of privatization are also where governments are weakest, where they are unable to provide the oversight and management functions necessary to protect public interests. This contradiction poses the greatest challenge for those who hope to make privatization work successfully.

Despite the vociferous, and often justified, opposition to water privatization, proposals for public–private partnerships in water supply and management are likely to become more numerous in the future. We do not argue here that privatization efforts must stop. We do, however, argue that all privatization agreements should meet certain standards and incorporate specific principles. Consequently, we offer the following Principles and Standards for privatization of water-supply systems and infrastructure.

Continue to Manage Water as a Social Good

Meet basic human needs for water

All residents in a service area should be guaranteed a basic water quantity under any privatization agreement. Contract agreements to provide water services in any region must ensure that basic human water needs are met first, before more water is provided to existing customers. Basic water requirements should be clearly defined (Gleick, 1996, 1999).

Meet basic ecosystem needs for water

Natural ecosystems should be guaranteed a basic water requirement under any privatization agreement. Basic water-supply protections for natural ecosystems must be put in place in every region of the world. Such protections should be written into every privatization agreement, enforced by government oversight.

The basic water requirement for users should be provided at subsidized rates when necessary for reasons of poverty

Subsidies should not be encouraged blindly, but some subsidies for specific groups of people or industries are occasionally justified. One example is subsidies for meeting basic water requirements when that minimum amount of water cannot be paid for due to poverty.

Use Sound Economics in Water Management

Water and water services should be provided at fair and reasonable rates

Provision of water and water services should not be free. Appropriate subsidies should be evaluated and discussed in public. Rates should be designed to encourage efficient and effective use of water.

Whenever possible, link proposed rate increases with agreed-upon improvements in service

Experience has shown that water users are often willing to pay for improvements in service when such improvements are designed with their participation and when improvements are actually delivered. Even when rate increases are primarily motivated by cost increases, linking the rate increase to improvements in service creates a performance incentive for the water supplier and increases the value of water and water services to users.

Subsidies, if necessary, should be economically and socially sound

Subsidies are not all equal from an economic point of view. For example, subsidies to low-income users that do not reduce the price of water are more appropriate than those that do because lower water prices encourage inefficient water use. Similarly, mechanisms should be instituted to regularly review and eliminate subsidies that no longer serve an appropriate social purpose.

Private companies should be required to demonstrate that new water-supply projects are less expensive than projects to improve water conservation and water-use efficiency before they are permitted to invest and raise water rates to repay the investment

Privatization agreements should not permit new supply projects unless such projects can be proven to be less costly than improving the efficiency of existing water distribution and use. When considered seriously, water-efficiency investments can earn an equal or higher rate of return to that earned by new water-supply investments. Rate structures should permit companies to earn a return on efficiency and conservation investments.

Maintain Strong Government Regulation and Oversight

Governments should retain or establish public ownership or control of water sources

The 'social good' dimensions of water cannot be fully protected if ownership of water sources is entirely private. Permanent and unequivocal public ownership of water sources gives the public the strongest single point of leverage in ensuring that an acceptable balance between social and economic concerns is achieved.

Public agencies and water-service providers should monitor water quality

Governments should define and enforce water-quality laws. Water suppliers cannot effectively regulate water quality. Although this point has been recognized in many privatization decisions, government water quality regulators are often underinformed and underfunded, leaving public decisions about water quality in private hands. Governments should define and enforce laws and regulations. Government agencies or independent watchdogs should monitor, and publish information on, water quality. Where governments are weak, formal and explicit mechanisms to protect water quality must be even stronger.

Contracts that lay out the responsibilities of each partner are a prerequisite for the success of any privatization

Contracts must protect the public interest; this requires provisions ensuring the quality of service and a regulatory regime that is transparent, accessible and accountable to the public. Good contracts will include explicit performance criteria and standards, with oversight by government regulatory agencies and NGOs.

Clear dispute-resolution procedures should be developed prior to privatization

Dispute resolution procedures should be specified clearly in contracts. It is necessary to develop practical procedures that build upon local institutions and practices, are free of corruption and difficult to circumvent.

Independent technical assistance and contract review should be standard

Weaker governments are most vulnerable to the risk of being forced into accepting weak contracts. Many of the problems associated with privatization have resulted from inadequate contract review or ambiguous contract language. In principle, many of these problems can be avoided by requiring advance independent technical and contract review.

Negotiations over privatization contracts should be open, transparent and include all affected stakeholders

Numerous political and financial problems for water customers and private companies have resulted from arrangements that were perceived as corrupt or not in the best interests of the public. Stakeholder participation is widely recognized as the best way of avoiding these problems. Broad participation by affected parties ensures that diverse values and varying viewpoints are articulated and incorporated into the process. It also provides a sense of ownership and stewardship over the process and resulting decisions.

We recommend the creation of public advisory committees with broad community representation to advise governments proposing privatization; formal public review of contracts in advance of signing agreements; and public education efforts in advance of any transfer of public responsibilities to private companies. International agency or charitable foundation funding of technical support to these committees should be provided.

Conclusions

As the 21st century unfolds, complex and new ideas will be tested, modified and put in place to oversee the world's growing economic, cultural and political connections. One of the most powerful and controversial will be new ways of managing the global economy. Even in the first years of the new century, political conflict over the new economy has been front and center in the world's attention.

This controversy extends to how fresh water is to be obtained, managed and provided to the world's people. In the water community, the concept of water as an 'economic good' has become the focal point of contention. In the last decade, the idea that fresh water should be increasingly subject to the rules and power of markets, prices and international trading regimes has been put into practice in dozens of ways, in hundreds of places, affecting millions of people. Prices have been set for water previously provided for free. Private corporations are taking control of the management, operation and sometimes even the ownership of previously public water systems. Sales of bottled water are booming. Proposals have been floated to transfer large quantities of fresh water across international borders, and even across oceans. These ideas and trends have generated enormous controversy. In some places and in some circumstances, treating water as an economic good can offer major advantages in the battle to provide every human with their basic water requirements, while protecting natural ecosystems.

Letting private companies take responsibility for managing some aspects of water services has the potential to help millions of poor receive access to basic water services. But in the past decade, the trend toward privatization of water has greatly accelerated, with both successes and spectacular failures. Insufficient effort has been made to understand the risks and limitations of water privatization, and to put in place guiding principles and standards to govern privatization efforts.

There is little doubt that the headlong rush toward private markets has failed to address some of the most important issues and concerns about water. In particular, water has vital social, cultural and ecological roles to play that cannot be protected by purely market forces. In addition, certain management goals and social values require direct and strong government support and protection, yet privatization efforts are increasing rapidly in regions where strong governments do not exist. We strongly recommend that any efforts to privatize or commodify water be accompanied by formal guarantees to respect certain principles and support specific social objectives. Among these are the need to provide for the basic water needs of humans and ecosystems as a top priority. Also important is ensuring independent monitoring and enforcement of water quality standards, equitable access to water for poor populations, inclusion of all affected parties in decision making and increased reliance on water-use efficiency and productivity improvements.

Openness, transparency and strong public regulatory oversight are fundamental requirements in any efforts to share the public responsibility for providing clean water to private entities.

Water is both an economic and social good. As a result, unregulated market forces can never completely and equitably satisfy social objectives. Given the legitimate concerns about the risks of this 'new economy of water', efforts to capture the benefits of the private sector must be balanced with efforts to address its flaws. Water is far too important to the well-being of humans and our environment to be placed entirely in the private sector.

Annex 4.1

Stakeholders Categories

Stakeholders are generally identified in terms of their interests in the issue addressed, their common perceptions of the values involved and their potential to affect the particular issue. Stakeholder groups usually vary in structure in terms of their potential to influence policy. As all these groups can affect the success of policy implementation and because governmental decision makers should represent both the powerful and disenfranchised equally, all stakeholders should be included in the assessment and search for policy resolution. Several different stakeholder categories are likely to be relevant in the assessment of a given issue. Among the criteria for recognizing these categories are ethnicity, gender, socio-economic status, religion, income source, land-use type and membership in existing institutions such as castes, unions, churches and cooperative groups. Stakeholder groups should be identified on the basis of whether there are common factors that are associated with them that affect their attitudes and the behaviour of those who are directly involved in the issue in question. Stakeholder categories that are relevant to dealing with land, water and other natural resources issues are presented below (Gregersen *et al.*, 1994a,b). Note that within each category there can be a range of interactions such as cooperation, competition and conflict among different stakeholder groups for the same natural resources. It should be emphasized that the people who fit into a particular category of stakeholders can also fit into one or more of the other categories. It, therefore, is important that policy makers understand and appreciate how these groups interact within and across the different stakeholder groups considered.

Watershed Dwellers and People Dependent on Watershed Resources

This group of stakeholders includes hunters and gathers, shifting cultivators and pastoralists who directly depend on the natural resources found on watershed landscapes for their existence. Although these people often wish to maintain their individual cultures and traditional uses of the land, they also want to improve their livelihood security and household incomes. This group of stakeholders generally has limited power to change the policies affecting them. They are usually interested in sustaining a desired level of direct consumption and use of the goods and services – including food, fuel, shelter and medicines – that can be derived from the watershed. These stakeholders may value the land for its spiritual values and role in their culture.

Commercial Users of Watershed Resources

These stakeholders include private or public groups or organizations that harvest products on watershed landscapes and, in most instances, are concerned with the economic consequences of these uses to themselves and to others. They have – or possibly could have – interest in sustaining the productivity of the watershed through appropriate property entitlements, although they might also be inclined to encourage such actions as species conversions. These stakeholders generally include local, regional, national and/or international manufacturers and markets of commodity products, tourism services and related activities.

Small-scale Converters of Watershed Landscapes

Members of this group of stakeholders are generally land-poor peasants who convert forests and other wildland ecosystems found on watershed landscapes to agricultural uses. These people have limited alternatives and commonly practice slash-and-burn agriculture and livestock grazing or sedentary agriculture production when rainfall is sufficient. They are often at risk of – or are already – degrading the lands that they have moved onto. These stakeholders frequently share the perception that they have only limited alternatives to sustain themselves. They are sometimes given forest lands for their use in resettlement schemes and at other times they are illegal encroachers who have little or no land or resource security and, therefore, limited incentive to think in the long term.

Large-scale Converters of Watershed Landscapes

These stakeholders include individuals, enterprise groups and government agencies with large commercial interest in the areas that underlie forest land and seek to convert forest land to other commercial uses. This group of stakeholders consists of ranchers, miners, urban and suburban developers and those people who construct roads, dams, power lines and other mostly public facilities that require the conversion of forest land to non-forest uses. Therefore, their interests are not in forests as such but in the land that is occupied by forests. In many cases, forests are viewed by these people as impediments to other uses that are valued more highly by this group. The alternative non-forest uses implemented by these people might or might not cause land and/or natural resources degradation and such degradation – if it might take place – may or may not be a concern to these people. These stakeholders generally have power to negotiate interests and influence policy.

Downstream Land Users

These stakeholders include those who are significantly affected by the land-use practices and other activities on upstream watersheds. This group might include some of people classified into other stakeholder groups, such as downstream irrigators, owners of hydropower dams, people in municipalities that use water for public distribution systems and downstream farmers. These stakeholders are often not considered in planning and implementing watershed management practices. However, on the other hand, it can be appropriate to internalize the positive downstream impacts of upstream watershed management practices and, in doing so, make appropriate compensation to upstream residents. In both of these cases, there are groups of people that have 'worked out' the necessary relationships between upstream and downstream land users and owners. Some of the downstream land users might have considerable power to influence policy, while other might have virtually none.

Consumers of Goods and Service Derived from Watershed Landscapes

Members of this stakeholder group are those who use and/or consume the variety of goods and services obtained from watershed landscapes. The group also includes those who personally use watersheds for recreation, scientific research, nature study and other relative activities. The interest of people in this group is related to the future availability and costs of the goods and services that they may want. These stakeholders have significant power in the marketplace through changes in their tastes and preferences, boycotts, etc. They include many of the most powerful interests in any society.

Environmental Advocacy Groups

These stakeholders are concerned with the local, regional, national and global environmental consequences of how watersheds are used, frequently seeking to improve and expand the protection and preservation of the natural resources on these landscapes. Included among their concerns are the loss of biodiversity, the preservation of rare and endangered species and ecosystems, global warming and air and water pollution. Some of these people might be concerned about environmental issues because of aesthetic and/or ethical issues, others because of impaired environmental service functions of watersheds, and still others because of potential impacts on the welfare of future generations. This group can have substantial power that is often linked to international agencies with lobbying power.

Social Development Groups

These stakeholders include government agencies, NGOs and other groups concerned with improving the social and economic well-being of rural and urban poor and other disadvantaged people. People within these agencies and organizations often have a 'strong interest' policies and programmes in land, water and other natural resources that can serve as a means of improving the lives of people. Use of the natural resources on watershed landscapes is simply one means of achieving people-oriented goals and objectives.

Government Agencies

Government agencies and personnel in these agencies should probably not be *stakeholders* in the traditional sense of the word. Instead, they should ideally represent all stakeholders including the voiceless ones such as future generations. In reality, however, these agencies and individuals within these agencies can have special interests in watershed management practices. Government officials often establish a *stake* in the natural resources that they deal with in executing their responsibilities. They might purchase tracts of land or become involved in industries that they deal with as a means to supplement inadequate incomes or to provide security for the future. Though these actions might be legal and ethical, they afford those who become involved in these actions a stake in the outcome of policy decisions and implementation. Beyond these more obvious situations, government agencies frequently act in a manner that fulfill their primary mission as they perceive it rather than being impartial arbiters of varied stakeholder interests. In this context, it must be recognized that different government agencies have their own perceptions of their respective missions and those missions can be conflicting rather than complementary with each other. All too often the resolution of these conflicts is based on political considerations irrelevant to the issues being decided. Additionally, it should be recognized that officials within a given agency – particularly those at different levels of responsibility within the agency such as field-level personnel, middle-level managers and high-level decision makers – can have different perceptions of their mission and, as a result, behave as different stakeholders. Needless to say, government agencies have *great power* to influence policy formulation and implementation in their inevitable though perhaps unintended capacity as stakeholders.

Bilateral, Multinational, and Non-governmental Organizations

The array of donors, funders or otherwise supporters of activities on watershed landscapes include a diverse group of local, regional, national and international organizations with wide-ranging interests in social and economic development and the sustainable use of natural resources. These organizations become stakeholders for a variety of political, social and religious reasons. Their interests are often expressed through *conditionalities* they place on loans, grants and other forms of support. On the one hand, local-level, regional and national NGOs play increasingly important roles in environmental advocacy, social action and community development programmes. They often offer services to rural communities that are sometimes beyond the capacity of the *central government* to provide. On the other hand, bilateral, multinational and international NOGs can bring external resources and occasional political pressure into the political arena. In recent years, NGOs at all levels are assuming increasingly stronger roles in the national and international policy arena as evidenced by the their influence in the 1992 international UNCED conference.

Bibliography

Gregersen, H., Brooks, K., Ffolliott, P., Lundgren, A. *et al.* (1994a) *Assessing Natural Resources Policy Issues.* Draft Policy Brief, EPAT/MUCIA/USAID, University of Minnesota, St. Paul, Minnesota.

Gregersen, H., Brooks, K., Ffolliott, P., Lundgren, A. *et al.* (1994b) *Assessing Natural Resources Policy Issues: A Framework for Developing Options.* Draft Policy Paper, EPAT/ MUCIA/USAID, University of Minnesota, St. Paul, Minnesota.

Annex 4.2

Time Value of Money, Discount Rates, Discounting and Compounding[1]

As mentioned in the text, IWM programmes generally involve a long-time horizon, with costs occurring over long periods and benefits from the programmes often not showing up for years after the expenses have been incurred and then also occurring over long periods. The basic idea of economics dictates that a *good* project is one where the benefits are at least as great as the costs, and preferably greater. But how can one compare costs and benefits when they occur at widely differing times?

Adding up costs and adding up benefits and then comparing them just does not work. The logic of this can be illustrated by a simple example. A completely trust-worthy acquaintance comes up to you and says, 'Lend me $10,000 today and I'll give it back to you 10 years from now.' If you are like 99.9% of the people in the world – and putting aside altruism and loyalty – you would say no. As another example, would you put money in a bank that pays zero interest? No. Among other things, you could get say 4% by putting it in your friendly savings and loan. There is, as the economist says, an *opportunity cost* involved in foregoing money now. If you lend the money to your acquaintance, you forgo the $400 a year that you could get in interest (4% interest on $10,000) by putting it in the bank. The basic point is that a $10,000 in hand today is worth more than $10,000 you would get 10 years from now, other things being equal. How much more is it worth? Well, if you estimate the level of interest rate that would get you to save and put money in a bank instead of spending it right now, you would get some idea of how much more it is worth. Thus, if 4% interest would get you to save another $100 of your income instead of spending it, then the $100 additional spending today is in you mind just about equivalent to waiting a year and getting $104 next year. Now if you left the $10,000 in the bank for 10 years instead of lending it to your aquaintance, you would have $14,802 in the bank 10 years from now.

Now how did we get that amount of $14,802? We get it using the simple *compound interest* formula. Thus, future value is determined by:

$$V_f = V_p(1 + r)^t \tag{1}$$

where V_f = future value; V_p = present value; r = *annual* compounding rate, or interest rate expressed as a decimal (4% = 0.04) and t = time, from year 1 through year n. Thus, $14,802 = 10,000 (1 + 4)^{10}$ This is called *compounding*.

Similarly, let's suppose we wanted to have $50,000 available to us 10 years from now. How much would we have to put in the bank today at 4% interest to get the $50,000 in 10 years? We use what is called the *discount* formula, which you can see is derived directly from the compounding formula above:

[1] Adapted in parts from Brooks *et al.*, 2003.

$$V_p = \frac{V_f}{(1 + r)^t} \tag{2}$$

For the above example, $\$50,000/(1.04)^{10} = \$33,780$. That is how much money we would need to put in the bank to have $\$50,000$ waiting for us 10 years from now. This is called *discounting*. Let's suppose that we find a place where we could earn 8% instead of 4%. How much would we then have to invest today to have the $\$50,000$ 10 years from now? The answer, using our formula above, is that we would only have to invest $\$21,810$ today. In other words, the higher the interest rate the greater the difference between equivalent values in different time periods.

The interest rate we could earn is also called the *opportunity cost of capital*. In other words if we pick the right interest rate, then by compounding and discounting, we can bring all values that take place in different time periods to a common point in time. Then we can legitimately compare the sum of costs with the sum of benefits for a project or activity that produces benefits; and we can see whether benefits are greater than or at least equal to the costs. If they are, then we can say that the investment appears to be an *economically efficient* use of funds, other things being equal and so long as we have used the right interest rate and so long as we have remembered the principle mentioned in the text: an investment is economically efficient if the total benefits are at least equal to costs and *if the benefits of each separable component of the project also has benefits that are at least equal to costs.*

So what is the right interest rate to use? In general it is the opportunity cost of capital for the entity undertaking the investment. If it is the government that is doing the investment, the interest rate it has to pay on borrowing funds might be the right one; or if it has no debt, then the right rate might be whatever rate it could earn on the next best alternative use for the funds. In practice, this becomes a policy matter.

The above is just a simple overview of the process of discounting and compounding to get values that occur at different points in time to be *equivalent* in terms of their values in the context of the time preferences of the investing entity. There are all kinds of shortcuts available, for example, formulas for figuring present value when values occur annually forever into the future, when they occur daily or monthly and so forth. They are all derived, however, from the simple compounding formula given above (formula 1 above). For the interested reader, a good place to start is Gregersen and Contreras (1992).

We can take this discussion one step further to get to the idea of discounting as a means of equating payments (benefits or costs) occurring at different points in time. Assume we are dealing with two projects and can only pick one. Furthermore, assume that there is one cost right now of $\$50$ in both projects. The only other values involved are a net benefit of $\$180$ in 5 years for project A and a net benefit of $\$215$ in 10 years in project B. If time did not matter, we would pick B since the benefit is higher. But time does matter, so we need to compare the two benefits at some common point in time. To do this, we discount both values back to the present. We find that the present value of $\$215$, 10 years from now and at a discount rate of 8%, is $\$215/(1.08)^{10} = \$100$. The present value of $\$180$ 5 years from now is $\$180/(1.08)^5 = \$122$. So project A turns out to be the best investment if 8% is the relevant discount rate; it has a higher present value than B.

Annex 4.3

Selected Bibliography of Useful References Dealing with the Economics of Natural Resources and Integrated Water and Watershed Management

This Annex provides a range of useful references for those who wish to get further into the process of economic valuation and analysis of watershed management practices, projects and programmes. This is only the 'tip of the iceberg' in terms of documentation available to help one to do economic analyses. We strongly encourage the reader who is interested to start with an older reference, Gregersen *et al.* (1987), since it follows most closely the general theme of this book: it is the integration of the biophysical data and information with value measures that will lead to the strongest and most useful economic assessments. If the biophysical data are not strong, then no matter how sophisticated one gets with the economics part, the results will not be very credible and therefore not useful.

At the end we provide some useful web sites.

Bibliography

Brooks, K.N., Gregersen, H.M., Berglund, E.R. and Tayaa, M. (1982) Economic evaluation of watershed projects – an overview methodology and application. *Water Resources Bulletin* 18, 245–250.

Shuhuai, D., Zhihui, G., Gregersen, H.M., Brooks, K.N. and Ffolliott, P.F. (2001) Protecting Beijing's municipal water supply through watershed management: an economic assessment. *Journal of the American Water Resources Association* 37, 585–594.

Dixon, J.A. and Hufschmidt, M.M. (eds) (1986) *Economic Valuation Techniques for the Environment: A Case Study Workbook.* Johns Hopkins University Press, Baltimore, Maryland.

Dixon, J.A., Fallon Scura, L., Carpenter, R.A. and Sherman, P.B. (1994) *Economic Analysis of the Environmental Impacts.* Earthscan, London.

EFTEC and Environmental Futures Ltd. (2006a) Valuing Our Natural Environment. Final Report NR0103 for the UK Department for Environment, Food and Rural Affairs, 20 March 2006.

EFTEC and Environmental Futures Ltd. (2006b) Valuing Our Natural Environment. Final Report NR0103 Annexes for the UK Department for Environment, Food and Rural Affairs, 20 March 2006.

FAO (2006) *The New Generation of Watershed Management Programmes and Projects.* FAO Forestry Paper 150. Food and Agriculture Organization of the United Nations, Rome.

Freeman, A.M., III (1993) *The Measurement of Environmental Resource Values.* Resources for the Future, Washington, DC.

Gregersen, H., Brooks, K., Dixon, J. and Hamilton, L. (1987) *Guidelines for Economic Appraisal of Watershed Management Projects.* FAO Conservation Guide 16. Food and Agriculture Organization of the United Nations, Rome.

Gregersen, H. and Contreras, A. (1992) *Economic Assessment of Forestry Project Impacts.* FAO Forestry Paper 106 for the World Bank, UNEP and FAO. Food and Agriculture Organization of the United Nations, Rome.

Gregersen, H., Arnold, J.E.M., Lundgren, A. and Contreras-Hermosilla, A. (1999) *Valuing Forests: Context, Issues and Guidelines.* FAO Forestry Paper 127. FAO of the United Nations, Rome, in collaboration with EPAT/MUCIA, the World Bank, the United Nations Development Program. Available at: http://www.fao.org/docrep/008/v7395e/v7395e00.HTM

Pagiola, S., von Ritter, K., and Bishop, J. (2004) *Assessing the Economic Value of Ecosystem Conservation*. Environment Department Paper No.101. The World Bank, Washington, DC, in collaboration with The Nature Conservancy and IUCN—The World Conservation Union.

Pearce, D.W. and Barbier, E.B. (2000) *Blueprint for a Sustainable Economy*. Earthscan, London.

Pearce, D., Whittington, D., Georgiou, S. and James, D. (1994) *Project and Policy Appraisal: Integrating Economics and Environment*. Organization for Economic Cooperation and Development, Paris.

Winpenny, J.T. (1991) *Values for the Environment: A Guide to Economic Appraisal*. HMSO, London.

Useful web sites for IWM planning and economics

http://www.iied.org/NR/forestry/projects/water.html (web site for IIED's Project on developing markets for watershed protection services and improved livelihoods.

http://www.flowsonline.net/ (this web site also deals with news and publications on payments for watershed services)

http://www.iucn.org/themes/wani/publications.html (this is the web site for the IUCN's Water and Nature Initiative).

http://www.gwptoolbox.org/index.cfm (The Global Water Partnership ToolBox is a compendium of good practices related to the principles of IWRM presented under a structured reference framework. The ToolBox allows water related professionals, to discuss, analyse the various elements of the IWRM process and facilitates the prioritization of actions aimed at improving the water governance and management. The IWRM ToolBox comprises an organized collection of case studies submitted by external contributors which have been peer reviewed. Through this website the ToolBox aims to faciliate that professionals and specialists engage with a broader community for the solution of (water related) problems.) See also the GWP's general publications web site for many useful documents on IWM planning http://www.gwpforum.org/servlet/PSP?iNodeID=231&iFromNodeID=102

http://topics.developmentgateway.org/water (a good gateway to sources of information on water resources management and planning worldwide).

http://www.epa.gov/owow/watershed/publications.html (the USA Environmental Protection Agency page on watersheds and their publications related to IWM).

http://www.ecosystemvaluation.org/ (useful site for the non-economist interested in ecosystem valuation and related topics).

Annex 6.1

A Procedure for Monitoring Water Quality to Insure Credibility and Consistency

It often becomes necessary to depend on volunteers to work with professional personnel in collecting baseline data on water quality to compare with water quality standards for evaluating the effects of a watershed management practice on water quality constituents. Established procedures and methods should be followed in a monitoring programme of collecting this information to insure data credibility and consistency. One such procedure is that outlined by the EPA of the USA in their volunteer monitor's guidelines (EPA, 1996). The basis of the measurement and data acquisition elements of these guidelines are presented below.

Sampling Design

The sampling design for the monitoring programme including information on the types of samples required, sampling frequency, sampling period (interval) and how the sample sites will be selected must be specified. Potential constraints such as weather conditions, seasonal variation and streamflow and/or site access that might affect scheduled monitoring activities must also be indicated along with how these constraints will be handled. Explicit safety plans should also be included.

Sampling Methods

Water quality constituents (parameters) to be sampled, how the samples should be taken, the equipment and containers to be used, methods of sample preservation when necessary and holding times between the time of taking the samples and analysing them must be described. If the samples are to be composited (mixed), how this will be done must be clearly stated. Procedures for decontamination and cleaning of equipment must also be specified.

Sample Handling and Custody

Sample handling procedures are necessary when samples are bought from the field to a laboratory for analysis. These samples must be properly labelled in the field with a label indicating the sample location, the sample number, date, as well as the time of collection, sample type, sample preservation method and sampler's name. Included in these procedures should be a statement of how the samples will be delivered and/or shipped to the laboratory for analysis. A *chain-of custody* record to be kept by the responsible person (people) should follow the samples when collecting, transferring, storing, analysing and disposing of the samples.

Analytical Methods

The analytical methods and equipment needed for the analysis of each water quality constituent either in the field or the laboratory should be identified. If the monitoring programme is based on standard methods of water quality analysis, these methods must be known and cited. If the intended analytical methods are different than the standard methods, a published reference that describes the analytical methods to be used in the laboratory should be cited.

Quality Control

Measures of precisions, accuracy, representativeness, completeness, comparability and sensitivity help to evaluate possible sources of variability and error in the samples and, by doing so, increase the confidence placed on the results. Quality control, therefore, is crucial to the success of a monitoring programme. Quality control is achieved from a set of measures that are taken in the monitoring programme to identify and correct possible errors. These measures include the correct labelling of samples in the field, proper training of the laboratory analyst and equipment calibration, applying proper laboratory analysis and, analysis of samples with known concentrations of a water quality constituent and/or repeated analysis of the same sample.

Therefore, a number of quality control samples should be obtained in the monitoring programme to help identify when and/or how contamination might occur. For most monitoring programmes, there is no set number of quality control samples that should be taken. A general rule put forth by the EPA is that 10% of the samples should be quality control samples. That is, if 30 samples are collected, at least one additional sample must be added as a quality control sample.

Instrument and Equipment Testing, Inspection and Maintenance

Routine inspection and preventive maintenance of field and laboratory instruments, equipment and facilities should be scheduled. Necessary spare parts and replacement instruments and equipment should be readily available to prevent delays. Bottles, droppers and colour comparators should be checked and cleaned when necessary. Reagents are replaced annually and/or according to the manufacturer's recommendation. A log book should be kept to track scheduled maintenance of all instruments and equipments.

Instrument Calibration

The frequency of calibrating the sampling and analytical instruments and equipment should be stated. The standards or certified instruments and equipment to be employed in the calibration process must also be specified. Once obtained, how the calibration will be maintained should be indicated.

Inspection and Acceptance Requirements for Supplies

Whether the supplies such as sample bottles and reagents are adequate for the sustaining the monitoring programme should be determined on a continuing basis. Needed supplies, instruments and equipments should be purchased when necessary under the supervision of the responsible individual.

Data Acquisition

Any types of data or other information that are necessary in the monitoring programme but are not obtaining through the sampling process need to be identified. Examples of this type of information are the historical background and perspectives, information obtained from topographic maps or aerial photographs or reports from earlier and/or other monitoring activities. Limits of the interpretation and/or use of this information resulting from uncertainty about its quality should be stated.

Data Management

The path that the source data obtained in the monitoring programme from its field collection and laboratory analysis to storage and use should be clearly specified. Accuracy and completeness of the field and laboratory records must be checked and errors in calculations, data entry to databases and report writing minimized. Computer software and hardware used to manage the database should be stated (see next section) if a database management system is used.

Other protocols for monitoring water quality and, more generally, environmental monitoring are found in publications by the World Meteorological Organization (1988), Ward *et al.* (1990), Bartram and Ballance (1996), Biswas (1997), Harmancioglu (1999), Brooks *et al.* (2003) and Wiersma (2004).

Bibliography

Bartram, J. and Ballance, R. (eds) (1996) *Water Quality Monitoring: A Practical Guide to the Design and Implementation of Freshwater Quality Studies and Monitoring Programmes.* E & FN Spoon, London.

Biswas, A.K. (ed.) (1997) *Water Resources: Environmental Planning, Management, and Development.* McGraw-Hill, New York.

Brooks, K.N., Ffolliott, P.F., Gregersen, H.M. and DeBano, L.F. (2003) *Hydrology and the Management of Watersheds.* Iowa State University Press, Ames, Iowa.

Environmental Protection Agency (EPA) (1996) *The Volunteer Monitor's Guide to Quality Assurance Project Plants.* EPA 841-B-96-003, Office of Wetlands, Oceans and Watersheds, Environmental Protection Agency, Washington, DC.

Harmancioglu, N.B. (1999) *Water Quality Monitoring Network.* Kluwer Academic, Dordrecht, The Netherlands.

Ward, R.C., Loftis, J.C. and McBride, G.B. (1990) *Design of Water Quality Monitoring Systems.* Wiley, New York.

Wiersma, G.B. (ed.) (2004) *Environmental Monitoring.* CRC Press, Boca Raton, Florida.

World Meteorological Organization (1988) *Manual on Water-Quality Monitoring: Planning and Implementing of Sampling and Field Testing.* WMO Series 680, World Meteorological Organization, Geneva, Switzerland.

Annex 6.2

Database Management Systems and Database Models

Database management systems are systems and/or computer software to manage a database and execute operations on the data that are requested by the user. Database management systems have emerged as a common part of almost any information management activity. There are many types of database management systems ranging from small systems that can operate on personal (laptop) computers to large systems that run on mainframes. A general description of the characteristics of database management systems and a few of the database models are presented in this annex. However, a few definitions are presented before this discussion.

Definitions

- *Database* – a comprehensive collection of data records (see below) and other information arranged for easy retrieval and designed for applications for multiple users but not necessarily at the same time.
- *Database management system* – software necessary to use a database. It handles the storage, retrieval and update of records by allowing a user to sort through large amounts of information and display relationships that would be time-consuming and often impossible to generate in a manual system.
- *Field* – a group of characters (letters, numbers and/or symbols) that define a specific piece of data or other information.
- *Record* – a group of fields that are treated as one unit.
- *File* – a group of related records containing the same fields but not necessarily the same data or other information.
- *Key* – a field or combination of fields that are used to locate a specified record for interrogation.

General Description of Database Management Systems

Most database management systems are sets of software programs that control the organization, storage, retrieve and/or modify the data and/or other pieces of information in a database. Database management systems allow a user to display the information on a computer screen or terminal for ease of continual data entry and verification; to output selected information on the screen or printer; and/or to create new databases and/or append data from other sources such as spreadsheets or word processors. Database management systems accept requests for specified information from what is called an applications program that instructs the operating system to transfer the appropriate information.

Database management systems include:

- A modelling language to define the schema of each of the databases in the management system. The most common organizations of a database are the hierarchical, network and relational models (see below). Database management systems can be organized to provide for one, two or all three of these models. The most suitable data structure (fields, records and files) of a system depends largely on the applications and inquiries that will be made of the system.
- A structure that has been optimized to accommodate the large amounts of data and/or other information to be stored on a permanent data storage device.
- A database query language and report writer program allow a user to interrogate the database, analyse its content and update it whenever necessary. The security of the database management system is controlled by preventing unauthorized users from viewing and/or updating the database.
- A transaction mechanism to insure database integrity despite concurrent user accesses (concurrency control) and faults (fault tolerance). It also maintains the integrity of the data in the database by not allowing more than one user to update the same record at the same time. Duplicate records are prevented with uniquely defined index constraints.

Examples of Database Models

Database management systems differ widely from a technical standpoint. For example, the terms of hierarchical, network and relational refer to the way by which the information in a database management system is organized internally. The internal organization affects how quickly and flexibly a user can extract this information. Requests for information from a database are made in the form of a query from a user. The information that is obtained from a query of a database can be presented in a variety of ways. Most database management systems include a report writer program that enables a user to output the information in the form of a report. Many systems also include a graphics component to output the information in graphs and charts.

The most common organizations of a database are the hierarchical, network and relational models – listed in their order of general development and subsequent use. A brief description of these respective models is presented below, with further details on these models found in references on database management systems (Thuraisingham, 1997; Subrahmanian, 1998; Dittrich and Geppert, 2001).

A *hierarchical model* is organized into a tree-like structure. This structure allows for repeating information by using what are called *parent–child relationships*, with each parent having as many children as necessary but each child having only one parent. All of the attributes of a specific record are listed under an entity type that is equivalent to a table. Each record is represented as a row and an attribute as a column in the table. Entity types are related to each other by using 1/N mapping also know as *one to many* relationships. Hierarchical structures were widely used in the first mainframe database management systems but owing to their organizational restrictions, they are rarely related to the structures that are found in the real world and, therefore, are seldom applied today.

A *network model* is organized in a more flexible manner than a hierarchical model. Where the hierarchical model structures the data as a tree of records, the network model allows each record to have multiple parent and child records – forming a lattice structure. A primary argument in favour of the network model in comparison to the hierarchical model is that the network model allows a more natural modelling of relationships between entities. Although the network model was widely implemented as the organization of a database when it became available, it failed to become the dominant organization largely because the relational model offered a high-level and more declarative interface. As computer hardware became faster, the increased flexibility and productivity of the relational models pushed the network models out of extensive usage.

The basic assumption of a *relational model* is that all of the data are presented as mathematical *n*–ary relations, with an *n*–ary relation a subset of the Cartesian product of *n* domains. Reasoning about such data is done in two-valued predicate logic in the mathematical model, meaning that there are two possible evaluations for each proposition – either true or false. Parenthetically, some people think that predicate logic (which is inherently two-valued) is a key part of the rational method, while others think that a system that uses a form of three-valued logic including a third value such as unknown or not applicable can still be considered a relational model. Data in the relational model are operated upon by means of rational algebra or calculus, these being equivalent in expressive power. The relational model of data permits the designer of the database to create a consistent and logical representation of information. Consistency is achieved by including declared constraints in the database design – usually referred to as the logic schema. The theory behind this model includes a process of database normalization by which a design with specified properties can be selected from a set of logically equivalent alternatives.

Selecting Database Software

One should consider the following points in selecting the database software:

- The anticipated applications to be made of the database management system should be known – it is also important to remember, however, that these applications might change in the future;
- The software to be selected should be powerful enough to handle the more complicated task but one that will not be too difficult to use or take too long to learn;
- Expert personnel familiar with the selected software should be readily accessible to provide the necessary support system for resolving the myriad of problems that often arise in installing and operating software of any kind;
- The software to be selected should be compatible with the available computer hardware system;
- The budget limitations confronted – some software is commercially developed and, therefore, must likely have to be purchased, while other software is available in the public domain.

Bibliography

Dittrich, K.R. and Geppert, A. (eds) (2001) *Component Database Systems*. Morgan Kaufmann, San Francisco, California.
Subrahmanian, V.S. (1998) *Principles of Multimedia Database Systems*. Morgan Kaufmann, San Francisco, California.
Thuraisingham, B.M. (1997) *Data Management Systems: Evolution and Interoperation*. CRC Press, Boca Raton, Florida.
Vossen, G. (1991) *Data Models, Database Languages and Database Management Systems*. Addison-Wesley, Reading, Massachusetts.

Annex 6.3

Computer Simulation Models

Numerous plot studies and landscape-level experiments have provided information on the effects of watershed management practices and land-use changes on water resources and forest, rangeland, wildlife and other natural resources. While the information obtained from these investigations is invaluable and should be enlarged upon whenever possible, source data and other information that are obtained from one watershed can seldom be applied directly to estimate the impacts at other watersheds. Therefore, alternative means of estimating these impacts are required in many instances – one such means is through the application of computer simulation models. Computer simulation models are representations of actual systems that allow one to explain and, in many cases, predict the response of natural resources to watershed management practices and, in doing so, gain a better understanding of the impacts of the watershed management practices. These computer models are largely based on a systems approach to simulation, and differ in terms of how and to what extent each component of the ecosystem process is considered.

Computer simulation models are generally composites of mathematical relationships – some empirical and some based on theory. As one attempts to explain or predict the impacts of watershed management practices on increasingly complex systems, more detail and complexity are needed in the model formulation. However, the reality is that people's understanding of hydrologic and ecosystem functioning is not always sufficient enough to represent every process mathematically – this often leads to the development of models (or components of models) that are calibrated by fitting parameters and relationships to the conditions encountered. This calibration process involves adjusting parameters until the computed response that is simulated approximates the observed response. Once calibrated, models can be used to estimate the response of the ecosystem to new, independent input data.

Development of Computer Simulation Models

Computer simulation modelling is widely used in many fields of science and management. The primary reasons for this usage are the savings of time and costs and the flexibility of modelling as an analytical tool. Other reasons include the improved computer literacy of managers and researchers and the wide availability of personal (laptop) computers and software. The users of computer simulation models should never forget, however, that models are abstractions of actual systems and that the output from models is only an estimate of system response. Moreover, confidence limits of the predicted output values are difficult to ascertain.

Methods

Computer simulation models are developed in many ways. One approach is through regression analysis. It is best to use a regression model that expresses a natural relation between the variables in the curve-fitting process. Knowledge of the behaviour of variables employed in a relationship allows the selection of one specific regression model over another. This process often leads to the formulation of more detailed plot studies that help to better define cause-and-effect relationships. When cause-and-effect relationships cannot be identified, empirical relationships can be derived. Selection of the regression model to represent a particular set of data is somewhat of an art. The choice should be made with an awareness of the statistical properties of various regression models.

The assemblage of one or more appropriate predictive functions such as those defined by regression analyses can allow for the simulation of a natural ecosystem. The response of an ecosystem to different inputs and levels of inputs can be simulated with such models. The sensitivity of model outputs or response to changes in the value of regression constants (parameters) in the model can be determined and the effects of modifying the structure of the system on response can also be examined.

Furthermore, assessments of subsequent development of a system and requirements for additional support data can be determined.

Mathematical equations representing the predictive functions of a watershed management practice are initially assembled in a flow diagram in developing a computer simulation model. These functions and the necessary linkages are then translated into a set of instructions in a selected computer language. Next, the model components expressed in a computer language are entered into a computer along with the appropriate descriptive, or input, data. Finally, by executing the model, outputs that predict the response of the system to specified inputs are obtained. For ease of operation, many simulation models require input data introduced through answers to questions posed to the user by the computer program – these are termed interactive models, in contrast to assemblages of extensive data records that are input to batch models.

Desirable Characteristics

Characteristics generally desired in developing a computer simulation model to estimate the effects of watershed management practices on natural resources include:

1. The model should be activity oriented – it must be possible to represent the watershed management practice to be simulated and then be able to simulate the effects of the practice in the model proposed.
2. The model should be capable of simulating time effects – the effects of most watershed management practices can be expected to change in time. It is necessary, therefore, that time-dependent phenomena be appropriately represented in the model.
3. Spatial effects should be simulated – the spatial distribution of the effects of watershed management practices generally has an influence on the effects of these watershed management practices on natural resources. As a consequence, the model must be able to accept inputs that describe the spatial variability of both the watershed landscape and the management practices on the landscape.
4. Data availability should be considered – the computer simulation model should not include data requirements that are difficult, costly or time-consuming to collect or acquire from the watershed on which the model is proposed for use. The models must be able to operate on relatively extensive data bases that are readily available in many simulation situations.

Applications of Computer Simulation Models

A question frequently asked of a watershed manager is '...what are the anticipated effects and responses of natural resources to a proposed watershed management practice?' As suggested here, applications of computer simulation models can be useful in formulating the answer to this question in many instances. More specifically, computer simulation models can be applied to estimate the responses of natural resources to a watershed management practice. Included among these simulators are those that have been developed to estimate the following responses to management:

- Changes in streamflow, peak flows and low flows, and changes in the physical, chemical and bacteriological quality of surface water;

- Growth, yield and quality of the trees in a forest overstorey;
- Development, accumulation and spatial distribution of organic materials (tree leaves or needles, branchwood, etc.) on the soil surface, which (in turn) can be inputs to computer simulators of fuel loading characteristics for erosion and sedimentation rates, fire management, etc;
- Species composition, level of production and utilization of forage plants;
- Livestock and carrying capacities;
- Quality of habitats for wildlife species.

Criteria that computer simulation models should ideally meet to simulate the effects and responses of natural resources to a management practice include:

- Accuracy of prediction – it is desirable that the computer simulation models are developed in such a manner that the error-statistics are known. Those models with minimum bias and error variance are generally superior.
- Simplicity – it refers to the number of parameters that must be estimated and the ease with which the model can be explained to potential users.
- Consistency of parameter estimates – consistency of parameter estimates is an important consideration in the development of models that use parameters estimated by optimization techniques of some kind. Computer simulation models are likely to be unreliable if the optimum values of the parameters are sensitive to the period of record used in the simulation exercise, or if the values vary widely between similar ecosystems.
- Sensitivity of results to changes in parameter values – it is desirable that the models not be sensitive to input variables that are difficult to measure and costly to obtain.

These criteria are also useful in selecting a computer simulation model from the possible alternative models that might be available for a specified computer exercise (see below).

Selection Criteria

A large number of computer simulation models are available to assist a manager in predicting the effects of watershed management practices on natural resources. These models vary in terms of complexity, data requirements and other operational factors and in the type of information that can be provided through their execution. However, a computer simulation model can only be helpful to a natural resources manager if it meets the selection criteria including:

1. The purpose for which the model is to be used and its spatial and temporal resolutions, which helps to define the kind and accuracy of input information required; i.e. the information needed to operate the model.
2. The availability of input data and other information needed to operate the model.
3. The suitability of the model for the physical, biological and social conditions of the site to be modelled.
4. The time and economic constraints to operating the model.
5. The availability of appropriate hardware systems and, if necessary, supporting software programs.
6. The skills, experience and background of the user of the model.

Once a computer simulation model has been selected for application, it must be calibrated and tested for the conditions considered to determine its appropriateness for the situation. Even when a model has been successfully applied to conditions thought to be similar to that being considered, the model is likely to still require some calibration and testing because each site is unique.

Other Observations

Computer simulation techniques are often available in a variety of scales and spatial resolution options. However, informational needs and data availability are likely to vary from one area to the next, which makes one standard and inflexible computer simulation technique impractical. What is often needed in these situations is a framework in which individual hydrologic processes (interception, infiltration, surface runoff, etc.) and natural resource components (water, timber, forage, etc.) are represented by modules that can be linked together to meet specific simulation objectives. An advantage of this framework is that the modules can be updated or replaced as needed, without disrupting other simulators in the modular framework. Users of a modular systems of computer simulation models can add or delete modules to accommodate particular informational requirements, and can easily adjust the mix of modules from one situation to another. Besides being more flexible, the modular approach is less expensive to operate, easier to use and requires less data to operate than larger, all-purpose computer simulation models.

Selecting computer simulation models for specific applications involves a compromise between theory or completeness and practical considerations. Locally derived models can be based on limited data and be empirical in nature, limiting their application elsewhere. On the other hand, more complete and theoretical simulation models might be better suited for widespread application, but can require input data or other information for application that are not available. One can choose from a considerable array of software that has been developed and that incorporates most standard simulation methods. In cases where computer models are not available that meet particular needs, models must be developed. This development requires that one have the necessary functional relationships, which are based largely on theory, plot studies, landscape-level experiments and other (previous) research. The developmental process, therefore, can be circular, in that computer simulation models are often used when the results of plot studies are lacking, but information from plot studies generally form a basis for this work in those situations where simulation models are not available and must be constructed.

Bibliography

Feldman, A.D. (1981) Models for water resources simulation: theory and experience. *Advanced Hydrology* 12, 297–418.
Hann, C.T., Johnson, H.P. and Braaakensiek, D.L. (1982) *Hydrologic Modeling of Small Watersheds*. Monograph Number 5, American Society of Agricultural Engineers, St. Joseph, Michigan.

Larson, C.L. (1971) Using hydrologic models to predict the effects of watershed modifications. In: *National Symposium on Watersheds in Transition*. Colorado State University, Fort Collins, Colorado, pp. 113–117.

Roberts, N., Andersen, D., Deal, R., Garet, M. and Shaffer, W. (1983) *Introduction to Computer Simulation: The Systems Dynamics Approach*. Addison-Wesley, Reading, Massachusetts.

Annex 6.4

Geographic Information Systems

There has been an unparalleled evolution of computing technologies in recent years. GISs are examples of this accelerating development and are assuming a more important role as a tool in the planning and implementation of watershed management practices. These systems are largely designed to satisfy the recurring geographical information needs of managers and researchers. Therefore, GISs have been optimized to store, retrieve and update information on watersheds and ecosystems and they have been programmed to process this information upon demand in a format that meets the informational needs of their users.

Formats

One maintains a set of geographically registered data layers in a GIS for subsequent retrieval and analysis. These layers can be stored in either raster or vector form. In a raster (or cell-based) system, the layer is represented by an array of rectangular or square cells, each of which has an assigned value. The advantages of a raster system are:

- The geographic location of each cell is implied by its position in the cell matrix. The matrix can be stored in a corresponding array in a computer provided sufficient storage is available. Each cell, therefore, can be easily addressed in the computer according to its geographic locations.
- The geographic coordinates of the cells need not be stored since the geographic location is implied in the positions of the cells.
- Neighbouring locations are represented by neighbouring cells. Therefore, neighbouring relationships can be analysed conveniently.
- A raster system accommodates discrete and continuous data sets equally well and facilitates the intermixing of the two data types.
- Processing algorithms are simpler and easier to write than is the case in vector systems.
- Landscape boundaries are presented by different values. When these values change, the implied boundaries change.

Disadvantages of a raster system include:

- Storage requirements are larger than those for vector systems.
- The cell size determines the resolution at which the resource is represented. It is difficult to adequately represent linear features.

- Image access is often sequential – meaning that a user might have to process an entire map to change a single cell.
- Processing of descriptive data is more cumbersome than is the case with a vector system.
- Input data are mostly digitized in form. A user, therefore, must execute a vector-to-raster transformation to convert digitized data into a form appropriate for storage.
- It can be difficult to construct output maps from raster data.

In a vector (or line-based) system, the line work is presented by a set of connected points. A line segment between two points is considered a vector. The coordinates of the points are explicitly stored, and the connectedness is implied through the organization of the points in the database. Advantages of a vector system are:

- Much less storage is required than for raster systems.
- Original maps can be represented at their original resolutions.
- Streams, forests, roads and other resource features can be retrieved and processed individually.
- It is easier to associate a variety of resource data with a specified resource feature.
- Digitized maps need not be converted to a raster form.
- Stored data can be processed into raster-type maps without a raster-to-vector conversion.

Disadvantages of a vector system include:

- Locations of the vertex points need to be stored explicitly.
- The relationship of these points must be formalized in a topological structure that can be difficult to understand and manipulate.
- Algorithms for accomplishing functions that are the equivalent of those implemented on a raster system are more complex. Furthermore, the implementation can be unreliable.
- Continuously varying spatial data cannot be represented as vectors. A conversion to raster is required to process these types of data.

Sources of Error

The spatial data sets for a GIS are obtained from maps, aerial photographs, satellite imagery and traditional and global positioning surveys. Each source involves a number of steps and transformations from the original measurements to the final digital coordinates, each of which can represent an error in the system. The origin of common errors in a GIS include:

- Field measurements – all positional information ultimately relies on field measurements. These measurements can be relatively precise (such as those that define legal property boundaries) or they might only be approximate (such as 'eyeball locations' of inventory plots on topographic maps). Global positioning systems (GPSs) consisting of a control segment, a constellation of satellites, and GPS receivers provide lower costs and higher throughout than traditional surveying in many situations. However, GPSs are only appropriate for a limited number of data layers.

- Maps – manual and/or automated map digitization is currently the most common form of spatial data entry and, therefore, has the greatest impact on spatial accuracy of the data sets digitized. Manual digitizing involves putting a paper or mylar map on a digitizing surface and tracing the features to be entered. Original map accuracy is a key determinant of spatial data accuracy regardless of the digitizing method.
- Imagery – it is a common source of natural resource spatial data, for both initial database development and updates. Most imagery comes from aerial cameras and satellite scanners, although video cameras and airborne scanners are also becoming a popular means of obtaining necessary imagery. Aerial photographs have been routinely used in resource mapping for 50 years or more. Among the factors affecting the accuracy of photo-derived data layers include tilt and terrain distortion and lens or camera distortion. Automated classification of satellite imagery is an established method of land-cover mapping. Classification converts multiband reflectance data into a single-layer land-cover map which is registered to a geographic coordinate system. The accuracy of class-boundary location, therefore, is a function of classification accuracy, geometry of the image and quality of the registration.
- Digitization – positional accuracies during digitization are affected by the equipment and/or operator skill. Currently used digitizers possess accuracies and precisions of better than 0.0025 cm. However, errors of varying magnitudes (up to 15%) can still be observed for digitized arcs and polygons.
- Coordinate registration – this involves converting from the digitzer coordinates to the coordinate system of the map projection used for printing the source map. Positional errors can occur at any of the steps in the process, including identifying control points in both the geographic and digitizer space; choosing a mathematical transformation and estimating the coefficients; and applying the transformation to the digitized data in producing the output layer. While large blunders are easily detected, small or random errors are not. Control points must be obtained from ground surveys or – when field measurements are lacking – control is commonly digitized from geographic coordinate points drafted on the source map, for example, Universal Transverse Mercator (UTM) graticule intersections drafted on 1:24,000-scale base maps.

Applications

GISs might be one of the most important technologies that watershed managers have acquired in recent years. GISs, often used together with remote sensing, have a number of possible applications for these managers, including those in the following areas:

1. Inventorying and monitoring – GISs and remote sensing can play important roles in keeping current inventory information, including quantities of resources available, where they are located, and whether they are growing, shrinking or holding their own. Of inventorying and monitoring, the latter is becoming a major new thrust for watershed managers. Monitoring is crucial for two important reasons:

- Watershed management agencies often enter into agreements through court-ordered decrees, environmental impact decisions and purposeful planning that require the monitoring of impacts from specified management practices.

- Watershed managers should monitor the effectiveness of prescribed watershed management practices to obtain the feedback necessary to make corrections in the management practices when necessary.

2. Management planning – GISs can be a valuable tool in the planning of watershed management practices to be implemented within administrative boundaries and, as frequently is the case in planning, among ownerships because of the importance of larger spatial scales. No longer might a watershed manager only deal with one forest stand at a time without placing the management of the stand into a larger context. Watershed management planning in the absence of this larger context often results in unwanted and unexpected down sides, such as cumulative effects or ecosystem fragmentation.

3. Policy setting – with science a current focus of policy setting in watershed management issues, GISs offer a means of preparing and displaying information on watershed management alternatives for review by elected and appointed officials charged with policy considerations. Many of the maps available to watershed managers in the past had to be hand-drawn, a tedious, labour-intensive process introduction, no doubt, errors in transcription.

4. Research – both GISs and remote sensing can be critical in research efforts. Researchers often experience more problems in going into the field and conducting a series of replicated studies at the landscape-level than they do at smaller scales. With GISs, researchers are able to address the problems involving larger scales by synthesizing resource data, developing concepts and displaying the findings. Information about patches, edges, connectivity, cumulative effects and dispersed and aggregated activities can be included in this process.

5. Consensual decision making – GISs make possible interactive collaborations among managers, policy setters, researchers and stakeholders in the decision-making process. The time of watershed managers and researchers preparing and then presenting management alternatives to the public without incorporating current science, social perspectives and economic interests has passed. Today, professionals are participants and, at times, facilitators in the decision-making process, with possibilities of GIS playing a significant role in this regard.

Other Considerations

Other considerations in relation to the use of a GIS as a research tool, regardless of the system, are data organization, database functions, input, query and analysis, display, reporting, user interface and hardware. Data organization addresses to the raster versus vector issue. Database functions cover topics such as the use of operating system. Input includes digitizing and input from external sources. Overlay is a query and analysis topic. User interface is concerned with menus versus command modes and other methods of control. Display relates to graphics output, while reporting considers tabular reports. Hardware to the system observer is the most prominent consideration, but it is part of the environment for the experienced user.

Bibliography

Bolstad, P.V. and Smith, J.L. (1992) Errors in GIS. *Journal of Forestry* 90, 21–29.
Congalton, R.G. and Green, K. (1992) The ABCs of GIS. *Journal of Forestry* 90, 13–20.

Sample, V.A. (ed.) (1994) *Remote Sensing and GIS in Ecosystem Management.* Island Press, Covelo, California.

Star, J. and Estes, J. (1990) *Geographic Information Systems: An Introduction.* Prentice-Hall, Englewood Cliffs, New Jersey.

van Roessel, J.W. (1986) *Guidelines for Forestry Information Processing.* FAO Forestry Paper 74, Rome, Italy.

Annex 7.1

Establishing a Research Agenda

There are always informational gaps that can hinder the planning and implementation of an IWM practice. One can begin with a conceptual model of the larger watershed management programme to help in deciding what information is needed and/or what research should be undertaken to alleviate this problem in a particular situation. That

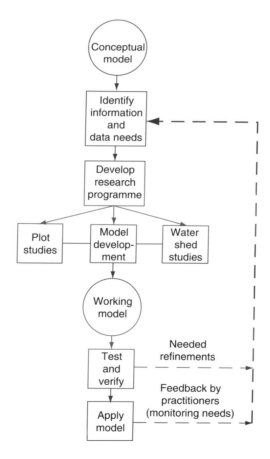

Annex Fig.7.1. Modelling approach in establishing a watershed research agenda. (Adapted from Brooks *et al.*, 1991.)

is, within the framework of the existing level of knowledge, a model that defines the system confronted and specifies the information needed as outputs can be developed. After such a model has been developed and the people have concurred in its structure and content, it becomes possible to identify what processes cannot be adequately quantified with the currently available information – this in turn points to the gaps in knowledge and, therefrom, a research agenda.

Once the research agenda has been satisfactorily established, the necessary investigations can be formulated and carried out. In some instances, these investigations might be a plot or small-scale process studies that are designed and implemented to furnish the source data and other information to better understand cause-and-effect relationships. Information obtained from these studies can be helpful in the development and testing of mathematical relationships that are needed for the synthesis of computer simulation models (see Annex 6.3). The results from these plot or small-scale process studies can also indicate a need to follow up with investigations on a larger-scale with watershed-level experiments, which are conducted to demonstrate and/or determined the hydrologic response – such as the magnitude, timing and quality of the water regimes – to a specified watershed practice (see Chapter 5).

Other possible themes or topics of a research agenda might be the development of improved mensurational protocols to measure hydrologic response; the enhancement of mapping and other spatial displays of source data; the synthesis of better spatial analysis techniques such as GISs (see Annex 6.4); the improvement of more efficient M&E programmes for biophysical and socio-economic systems; or the structuring of more efficiently accessed database management systems (see Annex 6.1). Research efforts such as these are often carried out by interdisciplinary teams of watershed researchers and managers to achieve optimal benefits.

Bibliography

Brooks, K. N., Ffolliott, P.F., Gregersen, H.M. and Thames, J.L. (1991) *Hydrology and the Management of Watersheds.* Iowa State University Press, Ames, Iowa.

Annex 7.2

Illustrative Modules and Sessions for Watershed Management Training Activities (Brooks and Ffolliott, 1993)

Module: Hydrologic process

- Session: Precipitation and interception;
- Session: ET and soil water;
- Session: Infiltration, runoff and streamflow.

Module: Plant–water relations

- Session: Energy relationships;
- Session: Plant water requirements and consumptive use;
- Session: ET estimates.

Module: Erosion processes and sedimentation

- Session: Erosion processes;
- Session: Prediction of soil loss;
- Session: Erosion control measures;
- Session: Sedimentation processes.

Module: Soil and water conservation

- Session: Mechanical and cultural methods of soil conservation;
- Session: Water harvesting and water spreading methods;
- Session: Water storage.

Module: Water quality and non-point pollution

- Session: Physical characteristics;
- Session: Dissolved chemical constituents;
- Session: Bacteriological quality;
- Session: Control measures.

Module: Integrative planning and management of larger river basins

- Session: Economic and environmental concerns;
- Session: Multi-purpose management considerations;
- Session: Linkages of upland conservation to larger river basins.

Module: Planning, monitoring, evaluation

- Session: The planning process;
- Session: Monitoring and evaluation;
- Session: Roles of computer simulation techniques (see following module).

Module: Simulation techniques, allocation of resources, decision making

- Session: Roles in sustainable development;
- Session: Computer models and simulation techniques;
- Session: Allocation procedures, economic appraisals and monitoring;
- Session: Decision-support systems.

Module: Policy considerations

- Session: Importance in sustaining the flows of natural resources;
- Session: Evaluation of the effectiveness of current (relevant) policies;
- Session: Policy measures.

Module: Environmental impact assessments

- Session: Types and scales of assessments;
- Session: Components of assessments;
- Session: Interpretations and applications of assessments.

Annex 7.3

Planning Watershed Management Training Activities

Careful planning is essential to a successful watershed management training activity regardless of the format to be followed. The first step in this planning process is a *needs assessment*. This step is largely a determination of who needs training, for what purpose and the type of activity that meets this purpose (Brooks and Ffolliott, 1993, 2002). This needs assessment ensures that the training fills a critical gap in the trainees knowledge or skills. In this step, one thinks about what training needs exist and the related knowledge that the trainees will bring to the activity. It is also important to think about the ways the organizers expect the trainees to apply the information after the training has ended.

Specifying Goals and Objectives

After identifying the trainee audience and otherwise completing the needs assessment, the goals and objectives of the training activity are specified, which also identifies who the trainees should be and the most appropriate training activity for them. To the extent possible, the targeted trainees should also help to determine the type of activity to be planned. However, if one chooses a particular training activity before-hand, such as a 1-day conference or a 2-week training course, the activity itself is likely to determine to a large extent the kind of participants who will attend. One should recognize and appreciate the trainees' understanding about the training subject to know where to start in terms of their understanding of the subject(s) and where to end the training activity. Depending on what this understanding is, a training activity could begin and end with achieving an awareness of the subject by providing information to promote the necessary awareness. This training is often presented at the strategic level.

A more detailed and extended training activity would provide facts and information to stimulate interest in the subject. Again, the training could be at the strategic level, although some tactical-level training might also be involved. An understanding of the subject is obtained by providing further knowledge so that the facts and information can be understood more thoroughly. This training is usually structured at the operational level. To bring the trainees to a stage of adaption and trial of the subject, training and practice is provided to enable the trainees to acquire skills on how to execute their assigned responsibilities better (Rogers and Shoemaker, 1971). A key step in the adoption process is moving the trainees through the crossover threshold from adoption and trail to adoption and implementation. The crossover threshold representing the trainee-change from knowing how to do something to the point that they widely adopt and implement the training topic (see Fig. 7.3) is reached at this point along the training continuum. Here, the activity would build on the earlier training base by promoting the use of the information presented earlier in practical situations. Adaption and implementation is achieved with this training at the operational level. Depending on what the responsibility of the trainers will be training can be a focus throughout.

Designing the Training Activity

After completing the needs assessment and identifying the trainees and training purposes, the type(s) of training activities, learning methods, as well as component modules and sessions to meet the goals and objectives are selected and the training activity is scheduled. Preparing necessary training materials is also a critical task. These materials must be technically correct, pertinent to reflect the region and be written in a language and presented at the educational and skill levels of the trainees. Training materials often include:

- A training guide that specifies the modules and sessions to be presented in the activity.
- A training manual that is the primary reference material provided the trainees and complements the learning methods to be used.
- Write-ups of case studies, examples and problem-solving exercises.
- Instructions for games and simulations, demonstration descriptions and field-trip itineraries.

Successful training requires efficient management. A poorly managed activity with confusion and delays can ruin the experience for everyone. Therefore, a training manager needs to be appointed with authority to do the job. A training manager should have good organizational skills and the necessary technical knowledge of the training focus. Only one person with specific and known tasks and responsibilities that are clearly understood by everyone should be appointed training manager. Adhering to the budget prepared for the training activity, getting equipment and materials and satisfying other needs are key responsibilities of the training manager. A contingency plan to meet a myriad of needs, problems and circumstances that might be encountered in any training activity helps in managing the activity.

Bibliography

Brooks, K.N. and Ffolliott, P.F. (1993) *Guidelines for Planning and Designing Training Activities in Forestry and Watershed Management.* Working Paper 12, Forestry for Sustainable Development Program, College of Natural Resources, University of Minnesota, St. Paul, Minnesota.

Brooks, K.N. and Ffolliott, P.F. (2002) International watershed management training activities. *Hydrological Science and Technology* 18, 55–64.

Rogers, E.M. and Shoemaker, F.F. (1971) *Communication of Innovations: A Cross Cultural Approach.* Free Press, New York.

References

Agus, F. and van Noordwijk, M. (eds) (2005) *Alternatives to Slash and Burn in Indonesia: Facilitating the Development of Agroforestry Systems. Phase 3 Synthesis and Summary Report*. World Agroforestry Center (ICRAF), Southeast Asia Regional Office, Bogor, Indonesia.

Altieri, M. (1988) *Environmentally Sound Small-scale Agricultural Projects: Guidelines for Planning*. CODEL-VITA Publication, Arlington, Virginia.

Anderson, H.W., Hoover, M.D. and Reinhart, K.G. (1976) *Forests and Water: Effects of Forest Management on Floods, Sedimentation, and Water Supply*. USDA Forest Service, General Technical Report PSW-18.

Andreassian, V. (2004) Waters and forests: from historical controversy to scientific debate. *Journal of Hydrology* 291, 1–27.

Baecher, G.B., Anderson, R., Britton, B., Brooks, K. and Gaudot, J. (2000) *The Nile Basin, Environmental Transboundary Opportunities and Constraints Analysis*. International Resources Group, Washington, DC for US Agency for International Development.

Baker, M.B., Jr (1986) Effects of ponderosa pine treatments on water yield in Arizona. *Water Resources Research* 22, 67–73.

Baker, M.B., Jr (comp) (1999) *History of Watershed Research in the Central Arizona Highlands*. USDA Forest Service, General Technical Report RMRS-GTR-29.

Baker, M.B., Jr, Ffolliott, P.F., DeBano, L.F. and Neary, D.G. (2004) *Riparian Areas of the Southwestern United States – Hydrology, Ecology and Management*. Lewis Publishers, Boca Raton, Florida.

Bell, J.W. (2000) The National forest road system: a public policy issue for the 21st century. In: Flug, M. and Frevert, D. (eds) *Watershed Management 2000: Science and Engineering Technology for the New Millennium*. American Society of Civil Engineers, Reston, Virginia. [CD-ROM] Windows.

Beschta, R.L. (2000) Watershed management in the Pacific Northwest: the historical legacy. In: *Land Stewardship in the 21st Century: The Contributions of Watershed Management*. Conference Proceedings, Tucson, Arizona, 13–16 March 2000. USDA Forest Service Proceedings RMRS-P-13.2000, pp. 109–116.

Blomquist, W., Schlager, E. and Heikkila, T. (2004) *Common Waters, Diverging Streams: Linking Institutions and Water Management in Arizona, California, and Colorado*. Resources for the Future, Washington, DC.

Bochet, J. (1983) *Management of Upland Watersheds: Participation of the Mountain Communities*. FAO Conservation Guide, Food and Agriculture Organization of the United Nations, Rome.

Bormann, B.T., Cunningham, P.G., Brookes, M.H., Manning, V.W. and Collopy, M.W. (1994) *Adaptive Ecosystem Management in the Pacific Northwest*. USDA Forest Service, General Technical Report PNW-GTR-341.

Bormann, B.T., Martin, J.R., Wagner, F.H., Wood, G.W., Alegria, J., Cunningham, P.G., Brookes, M.H., Friesema, P., Berg, J. and Henshaw, J.R. (1999) Adaptive management. In: Johnson, N.C., Malk, A.J., Sexton, W.T. and Sazro, R. (eds) *Ecological Stewardship: A Common Reference for Ecosystem Management*. Elsevier Science, Oxford, pp. 505–534.

Bosch, J.M. and Hewlett, J.D. (1982) A review of catchment experiments to determine the effects of vegetation changes on water yield and evapotranspiration. *Journal of Hydrology* 55, 3–23.

Brandes, D. and Wilcox, B.P. (2000) Evapotranspiration and soil moisture dynamics on semi-arid ponderosa pine hillslopes. *Journal of the American Water Resources Association* 36(5), 965–974.

Branson, F.A., Gifford, G.F., Renard, K.G. and Hadley, R.F. (1981) *Rangeland hydrology.* Kendall/Hunt Publishing, Dubuque, Iowa.

Brooks, K.N. and Ffolliott, P.F. (1993) *Guidelines for Planning and Designing Training Activities in Forestry and Watershed Management.* Working Paper 12, Forestry for Sustainable Development Program, College of Natural Resources, University of Minnesota, St. Paul, Minnesota.

Brooks, K.N. and Ffolliott, P.F. (2002) International watershed management training activities. *Hydrological Science and Technology* 18, 55–64.

Brooks, K.N., Gregersen, H.M., Lundgren, A.L., Quinn, R.M. and Rose, D.W. (1990) *Manual on Watershed Management Project Planning, Monitoring and Evaluation.* ASEAN-US Watershed Project, College, Laguna, Philippines.

Brooks, K.N., Gregersen, H.M., Ffolliott, P.F. and Tejwani, K.G. (1992) Watershed management: a key to sustainability. In: Sharma, N.P. (ed.) *Managing the World's Forests.* Kendall/Hunt Publishing, Dubuque, Iowa, pp. 455–487.

Brooks, K.N., Ffolliott, P.F., Gregersen, H.M. and Easter, K.W. (1994) *Policies for Sustainable Development: The Role of Watershed Management.* EPAT Policy Brief No. 6, Department of State, Washington, DC.

Brooks, K.N., Ffolliott, P.F., Gregersen, H.M. and DeBano, L.F. (2003) *Hydrology and the Management of Watersheds.* Iowa State Press, Ames, Iowa.

Brown, T.C. and Fogel, M.M. (1987) Use of streamflow increases from vegetation management in the Verde River Basin. *Water Resources Bulletin* 23, 1149–1160.

Brown, T.C., Brown, D. and Brinkley, D. (1993) Laws and programs for controlling nonpoint source pollution in forested areas. *Water Resources Bulletin* 22, 1–13.

Bruijnzeel, L.A. (1990) *Hydrology of Moist Tropical Forests and Effects of Conversion: A State of Knowledge Review.* UNESCO International Hydrological Programme, Paris, France.

Bruijnzeel, L.A. (2004) Hydrological functions of tropical forests: not seeing the soil for the trees? *Agriculture, Ecosystems and Environment* 104, 185–228.

Burchi, S. (1999) National regulations for groundwater: options, issues and best practices. Legal Papers Online #5. FAO of the United Nations, Rome. Available at: http://www.fao.org/Legal/prs-ol/full.htm

Burchi, S. (2005) The interface between customary and statutory water rights – a statutory perspective. Legal Papers Online #45. FAO of the United Nations, Rome.

Burdass, W.J. (1975) Water harvesting for livestock in Western Australia. In: Fraser, G.W. (ed.) *Proceedings of the Water Harvesting Symposium.* USDA Agricultural Research Service, ARS-W-22.

Bureau for the Near East (1993) *Water Resources Action Plan for the Near East.* US Agency for International Development, Washington, DC.

Calder, I.R. (2005) *Blue Revolution – Integrated Land and Water Resource Management*, 2nd edn. Earthscan, London.

Casley, D.J. and Kumar, K. (1987) *Project Monitoring and Evaluation in Agriculture.* Johns Hopkins University Press, Baltimore, Maryland.

Chandra, S. and Bhatia, K.K.S. (2000) Water and watershed management in India: policy issues and priority areas for future research. In: Ffolliott, P.F., Baker, B.M. Jr, Edminster, C.B., Dillon, M.C. and Mora, K.L. (tech. coords.) *Land Stewardship in the 21st Century: The Contributions of Watershed Management.* USDA Forest Service, Proceedings RMRS-P-13, pp. 158–165.

Chang, M. (2003) *Forest Hydrology: An Introduction to Water and Forests.* CRC Press, Boca Raton, Florida.

Clary, W.P. and Webster, B.F. (1990) Riparian grazing guidelines for the Intermountain regions. *Rangelands* 12, 209–212.

Cleverly, J.R., Dahm, C.N., Thibault, J.R., Gilroy, D.J. and Coonrod, J.E.A. (2002) Seasonal estimates of actual evapotranspiration from *Tamarix ramosissima* stands using three-dimensional eddy covariance. *Journal of Arid Environment* 52, 181–197.

Clyde, C.F., Israelsen, C.E. and Packer, P.E. (1976) Erosion during highway construction. In: *Manual of Erosion Control Principles and Practices*. Utah Water Research Laboratory, Utah State University, Logan, Utah.

Conant, F., Rogers, P., Baumgardner, M., McKell, C., Dasmann, R. and Reining, P. (1983) *Resource Inventory and Baseline Study Methods for Developing Countries*. American Association for the Advancement of Science, Washington, DC.

CWM/NRC (Committee on Watershed Management/National Research Council) (1999) *New Strategies for America's Watersheds*. National Academies Press, Washington, DC.

Davis, T.R.H. (1997) Using hydroscience and hydrotechnical engineering to reduce debris flow hazards. In: Chen, C. (ed.) *Debris-Flow Hazards Mitigation*. Proceedings, First International Conference American Society of Civil Engineers, New York.

DeBano, L.F. (1981) Water repellent soils: a state-of-the-art. USDA Forest Service, General Technical Report PSW-46.

DeBano, L.F., Neary, N.G. and Ffolliott, P.F. (1998) *Fire's Effects on Ecosystems*. Wiley, New York.

Dixon, J. and Wrathall, A. (1990) The reorganization of local government: reform or rhetoric? *New Zealand Journal of Geography* 9, 2–6.

Dunne, T. and Leopold, L.B. (1978) *Water in Environmental Planning*. W.H. Freeman, San Francisco, California.

ECE (Economic Commission for Europe) (2004) *Water-related Ecosystems: Features, Functions and the Need for a Holistic Approach to Ecosystem Protection and Restoration*. UN Economic and Social Council, document MP.WAT/SEM.4/2004/4 (8 October 2004). Paper prepared for meeting of the parties to the Convention on the Protection and Use of Transboundary Watercourses and International Lakes. Seminar on the Role of Ecosystems as Water Suppliers, Geneva, 13–14 December 2004. United Nations Economic Commission for Europe Secretariat, Geneva, Switzerland.

EcoSecurities (2006) EcoSecurities advises Pacific Hydro on the development of the first hydroelectric CDM project to be registered in Chile. Available at: http://www.ecosecurities.com/downloads/resource-139.pdf

EFTEC and Environmental Futures Ltd. (2006a) *Valuing Our Natural Environment*. Final Report NR0103 for the UK Department for Environment, Food and Rural Affairs, 20 March 2006.

EFTEC and Environmental Futures Ltd. (2006b) *Valuing Our Natural Environment*. Final Report NR0103 Annexes for the UK Department for Environment, Food and Rural Affairs, 20 March 2006.

Egan, A.F. (1999) Forest roads: where soil and water don't mix. *Journal of Forestry* 97, 18–22.

Environment Agency, UK (2006) *Water for Life and Livelihoods: A Framework for River Basin Planning in England and Wales*. Environment Agency, Almondsbury, Bristol, UK.

EPA (US Environmental Protection Agency) (1997) *Top 10 Watershed Lessons Learned*. Document EPA 840-F-97-001. EPA, Washington, DC.

Ethiopia Forestry Action Program (1993) *Volume I Executive Summary*, Final Report. Addis Ababa, Ethiopia.

FAO (1977) *Guidelines for Watershed Management*. FAO Conservation Guide 1. Food and Agricultural Organization of the United Nations, Rome.

Featherstone, J.P. (1996) Water resources coordination and planning at the federal level: the need for integration. *Water Resources Update* 104, 52–54.

Ffolliott, P.F. (1990) *Manual on Watershed Instrumentation and Measurements*. ASEAN-US Watershed Project, College, Laguna, Philippines.

Ffolliott, P.F. and Brooks, K.N. (1996) Process studies in forest hydrology: a worldwide review. In: Singh, V.P. and Kumar, B. (eds) *Surface-Water Hydrology*. Kluwer Academic, Amsterdam, The Netherlands, pp. 1–18.

Ffolliott, P.F. and Neary, D.G. (2003) Impacts of a historical wildfire on hydrologic processes: a case study in Arizona. In: *Watershed Management for Water Supply Systems: Proceedings of the International Congress*. American Water Resources Association, New York. [CD-ROM] Windows.

Ffolliott, P.F. and Thorud, D.B. (1977) Water yield improvement by vegetation management. *Water Resources Bulletin* 13, 563–571.

Ffolliott, P.F., Brooks, K.N., Gregersen, H.M. and Lundgren, A.L. (1995a) *Dryland Forestry: Planning and Management*. Wiley, New York.

Ffolliott, P.F., DeBano, L.F. and Ortega-Rubio, A. (1995b) Relationship of research to management in the Madrean Archipelago region. In: DeBano, L.F., Ffolliott, P.F., Ortega-Rubio, A., Gottfried, G.J., Hamre, R.H. and Edminster, C.B. (tech. coords.) *Biodiversity and Management of the Madrean Archipelago: The Sky Islands of Southwestern United States and Northwestern Mexico*. USDA Forest Service, General Technical Report GTR-RM-264, pp. 31–35.

Figallo, C. (1998) *Hosting Web Communities: Building Relationships, Increasing Customer Loyalty, and Maintaining a Competitive Edge*. Wiley, New York.

Fitzgibbon, J. (2006) Momentum building for saving streams. *The Oregonian*, 28 September 2006.

Gleick, P. (2002) *Dirty Water: Estimated Deaths from Water-Related Disease 2000–2020*. Pacific Institute Research Report. Pacific Institute for Studies in Development, Environment and Security, Oakland, California.

Gleick, P., Wolff, G., Chalecki, E. and Reyes, R. (2002) *The New Economy of Water: The Risks and Benefits of Globalization and Privatization of Fresh Water*. Pacific Institute for Studies in Development, Environment and Security, Oakland, California.

Gleick, P., Cooley, H. and Groves, D. (2005) *California Water 2030: An Efficient Future*. A Report of the Pacific Institute. Pacific Institute, Oakland, California.

Global Water Partnership (GWP/TAC) (2000) *Integrated Water Resource Management*. TAC Paper No. 4. Global Water Partnership, Stockholm. Available at: http://www.gwpforum.org/gwp/library/TACNO4.PDF

Gordon, A.M. and Newman, S.M. (1997) *Temperate Agroforestry Systems*. CAB International, Wallingford, UK.

Gordon, N.D., McMahon, T.A. and Finayson, B.L. (1992) *Stream Hydrology – An Introduction for Ecologists*. Wiley, New York.

Grace, J.M. III. (2000) Forest road side slopes and soil conservation techniques. *Journal of Soil and Water Conservation* 55, 96–101.

Gregersen, H. (1982) *Village Forestry Development in the Republic of Korea: A Case Study*. Document GCP/INT/347/SWE. Forestry for Local Community Development Programme, Food and Agriculture Organization of the United Nations, Rome.

Gregersen, H. and Contreras, A. (1992) *Economic Assessment of Forestry Project Impacts*. FAO Forestry Paper 106 for the World Bank, UNEP and FAO. Food and Agriculture Organization of the United Nations, Rome.

Gregersen, H.M., Brooks, K.N., Dixon, J.A. and Hamilton, L.S. (1987) *Guidelines for Economic Appraisal of Watershed Management Projects*. FAO Conservation Guide 16, Food and Agriculture Organization of the United Nations, Rome.

Gregersen, H., Brooks, K., Ffolliott, P., Lundren, A., Belcher, B., Eckman, K., Quinn, R., Ward, D., White, T., Josiah, S., Xu, Z. and Robinson, D. (1994) *Assessing Natural Resources Policy Issues*. EPAT/MUCIA/USAID Draft Policy Brief, University of Minnesota, St. Paul, Minnesota.

Gregersen, H., Easter, W.K. and deSteiguer, J.E. (2000) *Responding to Increased Needs and Demands*. In Land Stewardship in the 21st Century: The Contributions of Watershed

Management. Conference Proceedings. Proceedings RMRS-13. March 2000. U.S. Department of Agriculture, Rocky Mountain Research Station. Fort Collins, Colorado, p. 438.

Harr, R.D. (1983) Potential for augmenting water yield through forest practices in western Washington and western Oregon. *Water Resources Bulletin* 19(3), 383–393.

Haseltine, M., Baker, M.B., Jr and Hutchinson, B. (2002) Technology transfer of watershed management information. *Hydrological Science and Technology* 18, 77–88.

Helvey, J.D. and Patric, J.H. (1988) Research on interception losses and soil moisture relationships. In: Swank, W.T. and Crossley, D.A. (eds) *Forest Hydrology and Ecology at Coweeta*. Springer, New York, pp. 129–137.

Hem, J.D. (1992) *Study and Interpretation of the Chemical Characteristics of Natural Water*. US Geological Survey Water-Supply Paper 2254, Washington, DC.

Hewlett, J.D. and Troendle, C.A. (1975) Nonpoint and diffused water sources: a variable source area problem. In: *Proceedings of a Symposium on Watershed Management*, American Society of Civil Engineers, New York, pp. 21–46.

Hibbert, A.R. (1983) Water yield improvement potential by vegetation management in western rangelands. *Water Resources Bulletin* 19, 375–381.

Hodgson, S. (2003) *Legislation on Water Users' Organizations: A Comparative Analysis*. FAO Legislative Study, No. 79. FAO of the United Nations, Rome.

Hodgson, S. (2004) *Land and Water – The Rights Interface*. FAO Legislative Study 84. FAO of the United Nations, Rome.

Holechek, J.R., Pieper, R.D. and Herbel, C.H. (1998) *Range Management: Principles and Practices*. Prentice-Hall, Upper Saddle River, New Jersey.

Holling, C.S. (1978) *Adaptive Environmental Assessment and Management*. Wiley, Chichester, UK.

Horton, J.S. and Campbell, C.J. (1974) Management of phreatophyte and riparian vegetation for maximum multiple use values. USDA Forest Service Research Paper RM-117.

Huebner, D.P., Baker, M.B., Jr and Ffolliott, P.F. (2000) Increasing efficiency of information dissemination and collection through the World Wide Web. In: Ffolliott, P.F., Baker, M.B. Jr, Edminster, C.B., Dillon, M.C. and Mora, K.L. (tech. coords.). *Land Stewardship in the 21st Century: The Contributions of Watershed Management*. USDA Forest Service, Proceedings RMRS-P-13, pp. 420–423.

Ice, G.G. and Stednick, J.D. (eds) (2004) *A Century of Forest and Wildland Watershed Lessons*. Society of American Foresters, Bethesda, Maryland.

Ice, G.G., Adams, P.W., Beschta, R.L., Froehlich, H.A. and Brown, G.W. (2004) Forest management to meet water quality and fisheries objectives: watershed studies and assessment tools in the Pacific Northwest. In: Ice and Stednick (eds) *A Century of Forest and Wildland Watershed Lessons*. Society of American Foresters, Bethesda, Maryland, pp. 239–261.

Ingwersen, J.B. (1985) Fog drip, water yield, and timber harvesting in the Bull Run municipal watershed, Oregon. *Water Resources Bulletin* 21(3), 469–473.

Jacobs, L. (1986) *Environmentally Sound Small-scale Livestock Projects: Guidelines for Planning*. CODEL, New York.

Jane, A.M.J., Rennie, S.C. and Watkins, J.W. (2004) Integrated data management for environmental monitoring programs. In: Wiersma, G.B. (ed.) *Environmental Monitoring*. CRC Press, Boca Raton, Florida, pp. 37–62.

Janki, M. (2004) *Country Study on Customary Water Law and Practices in Guyana*. This paper was commissioned by IUCN under a joint FAO/IUCN research project investigating the interface of customary and statutory water rights, in progress.

Johnson, A. (2000) An effective method of utilizing the Internet to enhance watershed management. In: Flug, M. and Frevert, D. (eds) *Watershed 2000: Science and Engineering*

Technology for the New Millennium. American Society of Civil Engineers, Reston, Virginia. [CD-ROM] Windows.

Johnson, M.D. (2000) A sociocultural perspective on the development of U.S. natural resource partnerships in the 20th century. In: Ffolliott, P.F., Baker, M.B. Jr, Edminster, C.B., Dillion, M.C. and Mora, K.L. (tech. coords.) *Land Stewardship in the 21st Century: The Contributions of Watershed Management.* USDA Forest Service, Proceedings RMRS-P-13, pp. 205–212.

Johnson, N., Revenga, C. and Echeverria, J. (2001) Managing water for people and nature. *Science* 292, 1071–1072.

Jones, T. (2001) *Elements of Good Practice in Integrated River Basin Management: A Practical Resource for Implementing the EU Water Framework Directive.* Key issues, lessons learned and 'good practice' examples from the WWF/EC 'Water Seminar Series' 2000/2001, World Wide Fund for Nature, Brussels.

Karpiscak, M.M., Foster, K.E., Rawles, R.L., Wright, N.G. and Hataway, P. (1984) *Water Harvesting Agrisystem: An Alternative to Ground Water Use in the Avra Valley Area, Arizona.* Office of Arid Lands Studies, University of Arizona, Tucson, Arizona.

Kemper, K., Dinar, A. and Blomquist, W. (eds) (2006) *Institutional and Policy Analysis of River Basin Management Decentralization: The Principle of Managing Water Resources at the Lowest Appropriate Level – When and Why Does It (Not) Work in Practice?* The World Bank, Washington, DC.

Kessler, W.B., Salwasser, H., Cartwright, C.W. and Caplan, J.A. (1992) New perspectives for sustainable natural resources management. *Ecological Applications* 2, 221–225.

Kunkle, S., Johnson, W.S. and Flora, M. (1987) *Monitoring Stream Water Quality for Land-use Impacts: A Training Manual for Natural Resource Management Specialists.* Water Resources Division, National Park Service, Fort Collins, Colorado.

Kuruk, P. (2004) *Country Study on Customary Water Laws and Practices in Nigeria, 2004.* This paper was commissioned by the International Union for the Conservation of Nature (IUCN) under a joint FAO/IUCN research project investigating the interface of customary and statutory water rights, in progress.

Kusel, J., Doak, S.C., Carpenter, S. and Sturtevant, V.E. (1996) The role of the public in adaptive ecosystem management. In: *Sierra Nevada Ecosystem Project: Final Report to Congress: Volume II: Assessments and Scientific Basis for Management Options.* Centers for Water and Wildland Resources, University of California, Davis, California, pp. 611–624.

Lamb, J.C. (1985) *Water Quality and Its Control.* Wiley, New York.

Lant, C.L. (1999) Introduction – human dimensions of watershed management. *Journal of the American Water Resources Association* 35, 483–486.

Lawrence, S. and Giles, C.L. (1998) Searching the World Wide Web. *Science* 280, 98.

Lee, K.N. (1993) *Compass and Gyroscope: Integrating Science and Politics for the Environment.* Island Press, Washington, DC.

Lee, R. (1980) *Forest Hydrology.* Columbia University Press, New York.

Lee, S.W. (1981) Landslides in Taiwan. In: *Problems of Soil Erosion and Sedimentation.* Proceedings of the South-East Asian Regional Symposium, 27–29 January 1981, Bangkok, Thailand, pp. 195–206.

Leopold, L.B., Wolman, M.G. and Miller, J.P. (1964) Fluvial processes in geomorphology. Dover publications, New York.

Lu, S.Y., Cheng, J. and Brooks, K.N. (2001) Managing forests for watershed management in Taiwan. *Forest Ecology and Management* 143, 77–85.

MacDicken, K.G. and Vergara, N.T. (eds) (1990) *Agroforestry: Classification and Management.* Wiley, New York.

Magner, J.A. (2006) Total maximum daily loads, cohesive-sediment channel stability and water quality associated with geology and grazing in selected eastern Minnesota watersheds. PhD thesis, University of Minnesota, St. Paul, Minnesota.

Maidment, D.R. (ed.) (1993) *Handbook of Hydrology.* McGraw-Hill, New York.

Mastrull, D. (2006) Maps put towns in tricky waters: temple urged officials to use its data over FEMA's. Choosing either version will hurt someone. *Philadelphia Inquirer*, 1 October 2006. Available at: http://www.philly.com/mld/inquirer/news/local/15648304.htm

McLeod, G. (1990) Mixed grazing to the farmer in semi-arid Botswana. *Splash* 6, 15–16.

McNally, R. and Tognetti, S. (2002) *Tackling Poverty and Promoting Sustainable Development: Key Lessons for Integrated River Basin Management*. A WWF Discussion Paper. WWF-UK, Godalming, Surrey, UK.

Meehan, W.R. (1991) Influences of forest and rangeland management on salmonid fishes and their habitats. American Fisheries Society, Bethesda, Maryland.

Metz, A. (2000) Mideast's lifeblood: a region's water crisis. Available at: www.Newsday.com (25–27 June 2000).

Murray, C. and Marmorek, D. (2003) Adaptive management: a science-based approach to managing ecosystems in the face of uncertainty. *Proceedings of the Fifth International Conference on Science and Management of Protected Areas: Making Ecosystem Based Management Work*. Victoria, British Columbia.

Nair, P.K.R. (1989) *Agroforestry Systems in the Tropics*. Kluwer Academic, Dordrecht, The Netherlands.

National Research Council (1990) *Forestry Research: A Mandate for Change*. National Research Council, National Academy of Science, Washington, DC.

National Research Council (1999) *New Strategies for America's Watersheds*. National Research Council, National Academy of Science, Washington, DC.

National Research Council (NRC), (2000) *Watershed Management for Potable Water Supply – Assessing the New York City Strategy*. National Academy Press, Washington, DC.

Nowlan, L. (2004) *Customary Water Laws and Practices in Canada*. This paper was commissioned by FAO under a joint FAO/IUCN research project investigating the interface of customary and statutory water rights, in progress.

O'Connor, K.A. (1995) Watershed management planning: bringing the pieces together. MS thesis. California State Polytechnic University, Pomona, California.

OECD (2006) *Water: The OECD Experience*. Background paper prepared for the Fourth World Water Forum, Mexico City.

Olsen, M.E., Melber, B.D. and Merwin, D.J. (1981) A methodology for conducting social impact assessments using quality of life indicators. In: Finsterbusch, K. and Wolf, C.P. (eds) *Methodology of Social Impact Assessment*. Hutchinson Ross Publishing, Stroudsburg, Pennsylvania, pp. 43–78.

Pagiola, S. and Platais, G. (2002) *Payments for environmental services*. Environment Strategy Note 3. Environment Department, The World Bank, Washington, DC.

Peterman, R.M. and Peters, C.N. (1998) Decision analysis: taking uncertainties into account in forest resource management. In: Sit, V. and Taylor, B. (eds) *Statistical Methods for Adaptive Management Studies*. British Columbia Ministry of Forests, Victoria, British Columbia, pp. 105–127. ISBN 0-7726-3512-9.

Pilgrim, D.H., Boran, D.G., Rowbottom, I.A., Mackay, S.M. and Tjendana, J. (1982) Water balance and runoff characteristics of mature and cleared pine and eucalypt catchments at Lidsdale, New South Wales. In: O'Laughlin, E.M. and Bren, L.J. (eds) *First National Symposium on Forest Hydrology*. Melbourne, Australia, pp. 103–110.

Ponce, S.L. (1980) *Water Quality Monitoring Programs*. USDA Forest Service, Watershed Systems Development Group, USDG Technical Paper 00002.

Pratt, D.J. and Gwynne, M.D. (eds) (1977) *Rangeland Management and Ecology in East Africa*. Robert E. Krieger Publishing, Huntington, New York.

Quinn, R.M., Brooks, K.N., Ffolliott, P.F., Gregersen, H.M. and Lundgren, A.L. (1995) *Reducing Resource Degradation: Designing Policy for Effective Watershed Management*. EPAT Working Paper No. 22, Department of State, Washington, DC.

Raines Ward, D. (2002) *Water Wars: Drought, Flood, Folly and Politics of Thirst*. Riverhead Books, New York.

Reid, L.M. (1993) *Research and Cumulative Watershed Effects.* USDA Forest Service, General Technical Report PSW-GTR-141.

Rogers, E.M. and Shoemaker, F.F. (1971) *Communication of Innovations: A Cross Cultural Approach.* Free Press, New York.

Rogers, P. and Hall, A. (2003) *Effective Water Governance.* Global Water Partnership, Technical Committee. TEC Background Paper No. 7. Global Water Partnership, Sweden.

Rosgen, D. (1996) *Applied River Morphology.* Wildland Hydrology, Lakewood, Colorado.

Rosgen, D.L. (1994) A classification of natural rivers. *Catena* 22, 169–199.

Sarpong, G. (2004) *Going Down the Drain? Customary Water Law and Legislative Onslaught in Ghana.* This paper was commissioned by FAO under a joint FAO/IUCN research project investigating the interface of customary and statutory water rights, in progress.

Satterlund, D.R. and Adams, P.W. (1992) *Wildland Watershed Management.* Wiley, New York.

Sehlke, G. (2006) Introduction – adaptive management: what is it and where is it going? *Water Resources Impact* 8, 3–4.

Sengupta, S. (2006) In teeming India, water crisis means dry pipes and found sludge. *New York Times*, 29 September 2006.

Shank, B.M. (2002) The hydrologic characteristics of hybrid poplar plantations in contrast to natural forest stands, crops and fallow fields in northwestern Minnesota. MS thesis, University of Minnesota, St. Paul, Minnesota.

Sheng, T.C. (1990) *Watershed Conservation II.* A collection of papers for developing countries. The Chinese Soil and Water Conservation Society, Taipei, Taiwan, Republic of China and Colorado State University, Fort Collins, Colorado.

Shuhuai, D., Zhihui Geng, Gregersen, H.M., Brooks, K.N. and Ffolliott, P.F. (2001) Protecting Beijing's municipal water supply through watershed management: an economic assessment. *Journal of the American Water Resources Association* 37, 585–594.

Sidle, R.C. (2000) Watershed challenges for the 21st century: a global perspective for mountainous terrain. In: *Land Stewardship in the 21st Century: The Contributions of Watershed Management.* Conference Proceedings, Tucson, Arizona, 13–16 March 2000. USDA Forest Service Proceedings RMRS-P-13.2000, pp. 45–56.

Spedding, C.R.W. (1988) *An Introduction to Agricultural Systems.* Elsevier Applied Science, London.

St. Clair, T., Kurzbach, E.G., Truick, J., Knecht, G. and Boone, J.E. (2006) Adaptive management and the regulatory permitting process for water resource projects. *Water Resources Impact* 8, 14–17.

Stankey, G.H., Clark, R.N. and Bormann, B.T. (2005) *Adaptive Management of Natural Resources: Theory, Concepts, and Management Institutions.* USDA Forest Service, General Technical Report PNW-GTR-654.

Stednick, J.D., Troendle, C.A. and Ice, G.G. (2004) Lessons for watershed research in the future. In: Ice, G.G. and Stednick, J.D. (eds) *A Century of Forest and Wildland Watershed Lessons.* American Society of Foresters, Bethesda, Maryland, pp. 277–287.

Stoltenberg, C.H., Ware, K.D., Marty, R.J. and Wellons, J.D. (1970) *Planning Research for Resource Decisions.* The Iowa State University, Ames, Iowa.

Stouder, D.J., Bisson, P.A. and Naiman, R.J. (eds) (1997) *Pacific Salmon and Their Ecosystems: Status and Future Options.* Chapman & Hall, New York.

Thames, J.L. (1989) Water harvesting. In: *Role of Forestry in Combating Desertification.* FAO Conservation Guide, Rome, Italy, pp. 234–252.

Thomas, J.W. (2006) Adaptive management: what's it all about? *Water Resources Impact* 8(3), 5–7.

Tognetti, S., Mendoza, G., Aylward, B., Southgate, D. and Garcia, L. (2004) *A Knowledge and Assessment Guide to Support the Development of Payment Arrangements for Watershed Ecosystem Services (PWES).* Prepared for The World Bank Environment

---

Department with support from the Bank-Netherlands Watershed Partnership Program. Washington, D.C. Available at: http://www.flowsonline.net/data/pes_assmt_guide_en.pdf

Toupal, R.S. and Johnson, M. (1998) *Conservation Partnerships: Indicators of Success*. USDA Natural Resources Conservation Service, Social Science Institute Technical Report 7.1.

UNDP (2006) *Human Development Report 2006: Beyond Scarcity: Power, Poverty and the Global Water Crisis*. United Nations Development Programme, New York.

USDA, US Department of the Interior (1994) *Record of Decision for Amendments to Forest Service and Bureau of Land Management Planning Documents within the Range of the Northern Spotted Owl*. Bureau of Land Management, Portland, Oregon.

van Damme, H. (2001) Domestic water supply, hygiene, and sanitation. In: Meinnzen-Dick and Rosegrant (eds) *Overcoming Water Scarcity and Quality Constraints*. IFPRI 2020 Focus 9, Brief 3.

van der Leeden, F., Troise, F.L. and Todd, D.K. (1990) *The Water Encyclopedia*, 2nd edn. Lewis Publishers, Chelsea, Michigan.

Van Hylckama, T.E.A. (1970) Water use by saltcedar. *Water Resources Research* 6, 728–735.

Verry, E.S. (1986) Forest harvesting and water: the Lake States experience. *Water Resources Bulletin* 22, 1039–1047.

Verry, E.S., Hornbeck, J.W. and Dolloff, C.A. (eds) (2000) *Riparian Management in Forests of the Continental Eastern United States*. Lewis Publishers, Boca Raton, Florida.

Walters, C.J. (1986) *Adaptive Management of Renewable Resources*. Macmillian, New York.

Walters, C.J. and Holling, C.S. (1990) Large-scale management experiments and learning by doing. *Ecology* 71, 2060–2068.

Wetlands International (2003) *The Wetlands Initiative*. Annual Report 2003, Chicago, Illinois.

Whitehead, P.G. and Robinson, M. (1993) Experimental basin studies – an international and historical perspective of forest impacts. *Journal of Hydrology* 145, 217–230.

Wiersma, G.B. (ed.) (2004) *Environmental Monitoring*. CRC Press, Boca Raton, Florida.

Williams, J.R. (1975) Sediment yield predicted with universal equation using runoff energy factor. In: *Present and Prospective Technology for Predicting Sediment Yields and Resources*. USDA Agricultural Research Service, USDA-ARS-S-40, pp. 244–252.

Wischmeirer, W.H. (1975) Estimating the soil loss equation's cover and management factor for undisturbed areas. In: *Present and Prospective Technology for Predicting Sediment Yields and Resources*. USDA Agricultural Research Service, USDA-ARS-S-40, pp. 118–124.

Wolf, A. (2006) *Transboundary Fresh Water Dispute Database*. Oregon State University, Corvallis, Oregon. Available at: http://www.transboundarywaters.orst.edu/

Wolff, G. and Hallstein, E. (2005) *Beyond Privatization: Restructuring Water Systems to Improve Performance*. Pacific Institute, Oakland, California.

World Bank.n.d. (http://web.worldbank.org/WBSITE/EXTERNAL/TOPICS/ENVIRONMENT/ EXTEEI/0,contentMDK:20487921~menuPK:1187844~pagePK:210058~piPK:210062~the SitePK:408050,00.html)

World Wildlife Fund (WWF) (2003) Lessons from WWF's work for integrated river basin management. In: *Managing Rivers Wisely*. Available at: www.panda.org/livingwaters/publications

Zhang, H., Morison, J.I.L. and Simmondsm L.P. (1999) Transpiration and water relations of poplar trees growing close to the water table. *Tree Physiology* 19, 563–573.

Index